Beyond Biotechnology

Culture of the Land

A Series in the New Agrarianism

This series is devoted to the exploration and articulation of a new agrarianism that considers the health of habitats and human communities together. It demonstrates how agrarian insights and responsibilities can be worked out in diverse fields of learning and living: history, science, art, politics, economics, literature, philosophy, religion, urban planning, education, and public policy. Agrarianism is a comprehensive worldview that appreciates the intimate and practical connections that exist between humans and the earth. It stands as our most promising alternative to the unsustainable and destructive ways of current global, industrial, and consumer culture.

BEYOND BIOTECHNOLOGY

THE BARREN PROMISE OF GENETIC ENGINEERING

CRAIG HOLDREGE AND STEVE TALBOTT

THE UNIVERSITY PRESS OF KENTUCKY

Scholarly publisher for the Commonwealth,
serving Bellarmine University, Berea College, Centre College of Kentucky, Eastern Kentucky University, The Filson Historical Society, Georgetown College, Kentucky Historical Society, Kentucky State University, Morehead State University, Murray State University, Northern Kentucky University, Transylvania University, University of Kentucky, University of Louisville, and Western Kentucky University.

Editorial and Sales Offices: The University Press of Kentucky
663 South Limestone Street, Lexington, Kentucky 40508-4008
www.kentuckypress.com

All books in the Culture of the Land series are printed on acid-free recycled paper meeting the requirements of the American National Standard for Permanence in Paper for Printed Library Materials.

12 11 10 09 08 5 4 3 2 1

Index by Clive Pyne

Library of Congress Cataloging-in-Publication Data

Holdrege, Craig, 1953–
Beyond biotechnology : the Barren promise of genetic engineering / Craig Holdrege and Steve Talbott.
p. cm. — (Culture of the land)
Includes bibliographical references and index.
ISBN 978-0-8131-2484-1 (hardcover : alk. paper)
1. Genetic engineering. 2. Plant genetic engineering. 3. Agricultural biotechnology. I. Talbott, Steve. II. Title.
QH442.H63 2008
660.6'5—dc22 2007048491

Manufactured in the United States of America.

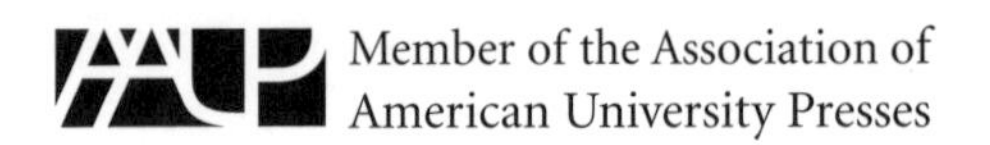

Contents

Illustrations

Preface

In Part I of this book we look at agricultural biotechnology. Our main concern is to show that we cannot understand genetic engineering and its implications unless we begin to view it within larger biological, organismic, ecological, economic, and societal contexts. Many of the problems of genetic engineering arise because we lack an awareness and understanding of these broader contexts. In fact, we live in illusions if we imagine genetic engineering as a way of making neat and discrete changes in organisms that contribute to just as neat and discrete programs for, say, solving the world's hunger problem. Without recognizing how our technical interventions are embedded within a complex web of relations, such "solutions" to problems cause even greater problems.

Genetic engineering is based on the premise that a gene is a clearly defined entity carrying out a specific function and, when transferred into a different organism, will perform the same function in the new context. That such manipulation often does not work according to plan can be viewed, theoretically, as a technical problem to be overcome. But it is actually a symptom of what scientists doing basic genetic research over the past decades have come to recognize: that the gene itself is context-dependent. This is the theme of Part II. The simple, straightforward gene that always does its job, oblivious to whether it is in a root cell or a leaf cell, a bacterium or a plant, an animal or a human being, does not actually exist. In fact, a good part of the "art" of genetic engineering entails limiting the implanted gene's responsiveness to its new and ever-changing context.

All genetic engineering is carried out with a specific living organism as the "medium" for gene expression. And, more generally, we speak of the human being, the dog, or the rose that "has" genes and consider these genes as fundamental to heredity and to the formation of the organism's characteristics. Through the single-minded focus on discovering the building blocks of heredity, the organism itself—as a whole, coherent being—recedes into the background. It becomes merely the carrier of genes or the result of genetic effects. This perception is one-sided and an artifact of the scientific, technological process that focuses on understanding and manipulating details (parts) as a way to effect changes in wholes. If you work in this way long enough you may forget what you are dealing with. But while the whole living organism may disappear from consciousness, it does not disappear from reality.

So in Part III we turn our attention to organisms. The discussions of the cow, the sloth, and (in Part IV) the skunk cabbage are attempts to portray, in a concrete way, the fundamental qualities of wholeness and integration that characterize each organism, but each in its own individual way. Each of these studies is only a beginning, but a beginning that is significant, since it entails a shift in awareness from "entities" to relations, qualities, and contexts—precisely the kind of awareness that is missing in what drives genetic engineering and, more broadly, our egocentric, utilitarian approach to the world.

When we make this shift, then the world no longer appears to us in the same way. In a subtle manner, more or less everything changes. We see ourselves, as we discuss in chapter 14, as participants in an ongoing, evolving conversation with the world. And every true conversation involves a back and forth between the conversing parties. So our fellow creatures, as partners in a conversation, elicit our interest, and we try to understand them from their perspective. There are no easy answers and no pat solutions to how such an interaction could or should play itself out. But it makes all the difference in the world that we become aware of our involvement in a participatory process, which calls on us to develop ever new capacities of perception, insight, and responsiveness. In this process we leave behind the all-too-comfortable stance of viewing and treating our fellow creatures as things to be manipulated.

The kind of shift we're speaking of is radical. To get at the roots of fundamental change you have to look at thought—deep-seated habits of thought that inform our every action. As David Bohm states, "The whole ecological problem is due to thought, because we have the thought that the world is there for us to exploit, that it is infinite, and so no matter what we did, the pollution would all get dissolved away. . . . Thought produces results, but thought says it didn't do it" (Bohm 1996, 11).

Since thought *does* do it, we take a careful look, especially in chapter 13, at some assumptions and habits of thought that inform modern science. These assumptions and thought-forms have opened up immense fields of inquiry. But at the same time they are sorely limited, and this limitation becomes problematic when contemporary scientific practice and theory become the standard for what counts as real knowledge. When this happens, other modes of inquiry are marginalized, and we don't even realize that immense fields of investigation lie fallow, because traditional scientific methodology ignores them. So while all science presupposes qualities, which form the fabric of human experience and inform our every action, it has given us no tools to better understand those qualities. Nonetheless, the application of science in technology qualitatively changes things in the world, but within the dominant paradigm we have no way to assess what we're doing.

It is not so hard to recognize that you can't solve problems with the same kind of thinking that causes the problems you're trying to address. It's another matter to change that thinking at its source. In Part IV we are concerned with an evolution of science—and more generally, human understanding—to encompass the qualities of the world. So while these epistemological considerations, taken in isolation, may seem distant from the pressing problems we discuss in the earlier chapters, they are not. Our social and environmental "fixes" will remain tenuous, always at risk of reversal, if we do not work to eradicate the problems at their source: in our view of the world and our habits of thought. After all, it required at least several hundred years for us to become the alienated culture we are today; we can only change direction through profound shifts in our most fundamental assumptions.

Part I

Genetic Engineering and Agriculture

Chapter 1

Sowing Technology

Drive the Nebraskan backroads in July, and you will encounter one of the great technological wonders of the modern world: thousands of acres of corn extending to the vanishing point in all directions across the table-flat landscape. It appears as lush and perfect a stand of vegetation as you will find anywhere on earth—almost every plant, millions of them, the same uniform height, the same deep shade of green, free of blemish, emerging straight and strong from clean, weed-free soil, with the cells of every plant bearing genetically engineered doom for the over-adventurous worm.

If you reflect on the sophisticated tools and techniques lying behind this achievement, you will likely feel some of the same awe that seizes so many of us when we see a jet airliner taking off. There can be no doubt about the magnitude of the technical accomplishment on those prairie expanses. And yet, the question we face with increasing urgency today is whether this remarkable cornucopia presents a picture of health and lawful bounty, or instead the hellish image of nature betrayed.

Actually, it is difficult to find much of nature in those cornfields. While nature always manifests itself ecologically—contextually—today's advanced crop production uproots the plant from anything like a natural, ecological setting. This, in fact, is the whole intention. Agricultural technology delivers, along with the seed, an entire artificial production environment designed to render the crop largely independent of local conditions. Commercial fertilizer substitutes for the natural fertility of the soil. Irrigation makes the plants relatively independent of the local climate. Insecticides prevent undesirable contact with local insects. Herbicides discourage social mixing with unsavory elements in the local plant population. And the crop itself is bred to be less sensitive to the local light rhythm.

Where, on the farm shaped by such technologies, do we find any recognition of the fundamental principle of ecology—namely, that every habitat is an intricately woven whole resisting overly ambitious efforts to carve it into separately disposable pieces?

But all this represents only one aspect of agriculture's abandonment of supporting environments. The modern agribusiness operation in its entirety has been wrenched free from the rural economic and social milieu that once sustained it. The farm itself is run more and more like a self-contained factory operation. And the trend toward vast monocultures—where entire ecologies of interrelated organisms are stripped down to a few, discrete elements—has become more radical step by step: first a single crop replacing a diversity of crops; then a single variety replacing a diversity of varieties; and now, monocultures erected upon single, genetically engineered traits.

As the whole process drives relentlessly forward, the organism itself becomes the denatured field in which genes are moved to and fro without regard to their jarring effect upon the living things that must endure them. Want to make a tobacco plant glow in the dark? Easy—inject a firefly gene! Want a frost-resistant strawberry? Try a gene or two from a cold-water flounder.

Yet, despite such chimera-like prodigies, the overriding question about biotechnology is not whether we are for or against this or that technical achievement, but whether the debate will be carried out in just such fragmented terms. In focusing on technological wonders to improve agriculture, are we losing sight of the things that matter most—the diverse, healthy, and complex communities and habitats we would like to live in? The question to ask of every technology is how it serves, or disrupts, the environment into which we import it.

Is Genetic Engineering New?

The natural setting whose integrity we need to consider first of all is that of the individual organism. The challenge we're up against here emerges in the frequently heard argument that genetic engineers are only doing what we've always done, but more efficiently. Writing in the *New York Times,* Carl B. Feldbaum, president of the Biotechnology Industry Organization, objected to the claim by critics that "what [traditional breeders] do is 'natural' while modern biology is not": "Archaeologists have documented twelve thousand years of agriculture throughout which farmers have genetically altered crops by selecting certain seeds from one harvest and using them to plant the next, a process that has led to enormous changes in the crops we grow and the food we eat. It is only in the past thirty years that we have become able to do it through biotechnology at high levels of predictability, precision and safety" (Feldbaum 1998).

But the concern about genetic engineering today isn't that it enables us to commit altogether new mistakes. Rather, it is that it perfects our ability to commit old ones. No one should suggest that the abuse of our technical powers began with the discovery of the double helix. Using conventional techniques, breeders have, for example, produced Belgian cattle with such overgrown muscles that they cannot be delivered naturally—birth requires Caesarian section. Likewise, there are hobbyist chicken breeders who—to judge from the pictures in their magazines—are more interested in bizarre effects that tickle human fancies than in the welfare of the chickens themselves.

The difference is that with genetic engineering we can now manipulate living organisms much more efficiently and more casually than ever before. The technician need scarcely be distracted by the animal itself. There's none of the Frankenstein drama and messiness. We can construct our monsters in a clean and well-lit place. The reassuring familiarity of the laboratory doubtless contributes to the illusion of precision and safety.

We begin to recognize the illusion when we observe how Feldbaum's claim completely glosses over what *is* unprecedented about genetic engineering: that it selects isolated genes, not entire healthy organisms. Writing in *Science*, geneticist Jon W. Gordon (1999) assesses the failed attempts to create heavier farm animals by inserting appropriate genes. In pigs, the addition of growth hormone–producing genes did not result in greater growth, but unexpectedly lowered body-fat levels. In cattle, a gene introduced to increase muscle mass "succeeded," but the growth was quickly followed by muscle degeneration and wasting. Unable to stand up, the experimental animal had to be killed. So much for precision.

Such results are hardly surprising when you consider the isolated and arbitrary intrusion represented by single-gene changes. By contrast—and this is what Feldbaum ignores—traditional breeding allows everything within the organism to change together in a coordinated way. As Gordon writes, "Swine selected [by traditional methods] for rapid growth may consume more food, produce more growth hormone, respond more briskly to endogenous growth hormone, divert proteins toward somatic growth, and possess skeletal anatomy that allows the animal to tolerate increased weight. Dozens or perhaps hundreds of genes may influence these traits."

If there's a logic to ecological relationships that says, "Change one thing and you change the whole," the same applies to the interior ecology of the organism. Responsible traditional breeding is a way of letting everything change without violating the whole—because it is the organism *as a coherent and healthy whole* that manages the change. Isn't it reasonable to assume that there's a wisdom at work amid all the complexity of the evolved organism that we cannot lay claim to with our largely trial-and-error manipulations?

Do Organisms Need Preserving?

In traditional breeding the integrity of the organisms themselves places limits upon what can be done—limits you could reasonably call "natural." For example, you could not cross a strawberry with a cold-water fish in order to obtain strawberries with "anti-freeze" genes.

The problem now is that we can break through these limits, but we have not replaced the safeguard they represented. Today, such a safeguard can come only from our own intimate, respectful understanding of the organism as a whole and of the ecological setting in which it exists.

This is the decisive question: does the organism possess a wholeness, an integrity, that demands our respect? And can we gain a deep enough understanding of it to say, "*This* change is a further expression of the organism's governing unity, and *that* change is a violation of it?" (See also Part III of this book.)

It is a difficult challenge, and not one we have trained ourselves to meet. You have to see a plant or animal in its own right and in its natural environment in order to begin grasping who or what it is. But given what ecologists David S. Wilcove at Environmental Defense and Thomas Eisner at Cornell University have called the "demise of natural history" in our time (Wilcove and Eisner 2000), there is not much hope of greater familiarity with the organisms whose natures we manipulate—certainly not by those laboratory- and test tube–bound researchers who are doing the manipulating.

Nevertheless, some things are fairly obvious. It's hard to understand how the mad cow debacle could have occurred if anyone had bothered to notice the cow. How could we possibly have fed animal parts to ruminants? *Everything* about the cow, from its teeth to its ruminating habits to its four-chambered stomach, fairly shouts at us, "*Herbivore!*" (See also chapter 10 on the cow.) Can we violate an organism's integrity in such a wholesale manner without producing disasters—for the organism, if not also for ourselves?

What the mad cow episode illustrates is that our notions of safety are relative to our understanding of the organism. And nothing has tended to fragment our view of the organism as powerfully as genetic engineering. Instead of a coherent whole expressing an organic unity through every aspect of its being, the engineers hand us a bag of separate traits and molecular instrumentation.

Are Bioengineered Products Adequately Tested?

Only such a fragmenting mentality could suggest (in the words of former U.S. Secretary of Agriculture Dan Glickman) that "test after rigorous scientific test have proven these [genetically engineered] products to be safe" (Glickman 1997). This suggestion is simply false on its face (see Smith 2007). The application to cows of bovine growth hormone (rBGH) produced by genetically engineered bacteria was approved primarily on the basis of tests with rats—not cows, and not people who consume cow products. Genetically altered Bt corn was approved without being tested for its effects on beneficial species such as green lacewings or on "incidental" species.

But the more fundamental problem is that, because the organism is an organic unity, its assimilation of foreign DNA potentially changes *everything*. Gene expression and protein levels are altered in ways that have proven consistently unpredictable. About 1 percent of genetic transfers yield the looked-for result; the other 99 percent are all over the map. For example, when scientists engineered tomatoes for increased carotene production, they indeed got some plants with more carotene—but those plants were unexpectedly dwarfed (Fray et al. 1995). No one expected this experiment to yield dwarfed plants.

So even the 1 percent statistic paints too optimistic a picture. This "success" rate reflects a focus on the particular trait that was looked for; but even when this trait is obtained and the resulting organism is used as the founding ancestor of a new, genetically altered line, it remains to ask: what about the subtle changes throughout the rest of the organism—changes not directly related to the researcher's intent? If there can be immediately obvious changes such as dwarfing, there can be many more unobvious ones. It's hard to test for changes when anything can happen and you don't know what you're looking for. In actual practice, almost no such testing is done.

Is Biotechnology Good for the Environment?

Against this backdrop, the biotech companies' promotion of genetically altered crops as the Great Green Hope of the environment due to the promise of reduced pesticide applications is puzzling at best. After all, the entire thrust of the factory-farmed monocultures encouraged by these companies is to eliminate across huge acreages all traces of any environmental richness that might have been worth preserving in the first place. And now the corporate research laboratories are poised to release into this devastated landscape a continuing stream of alien genes that, in their own right, promise to become the ultimate uncontrollable pollutants. Chemical spills can eventually be cleaned up, but there is no recalling the replicating genes we have loosed upon the natural world.

If there's any claim that must be evaluated ecologically, it's the claim of environmental benefit. Yet, as Michael Pollan remarks in a *New York Times Magazine* piece on genetically engineered potatoes, those who simply take vast monocultures for granted will always think they have, say, a Colorado potato beetle problem—rather than the total environmental problem of potato monoculture (Pollan 1998).

This detachment of particular problems from the environment as a whole invites a search for "silver bullets"—precise and complete fixes that can rid us of the problems. Certainly there are silver bullets to be had, even if their unfortunate tendency is to rip crudely through the delicate, ecological fabric they are aimed at. Perhaps the most obvious silver bullet is Bt cotton. The relatively mild Bt toxin engineered into the crop is highly effective against the bollworm and substitutes for an extraordinarily nasty series of sprayings in conventional cotton fields. Yet, to leave the matter there is to accept the conventional approach as the only alternative. And it is also, as Charles Benbrook (2001) points out, to be extremely irresponsible.

Benbrook is a former executive director of the National Academy of Sciences Board on Agriculture and now an agricultural consultant in Sandpoint, Idaho. He sees Bt, in its non–genetically engineered, externally applied form, as perhaps the most valuable pesticide ever developed. A naturally produced substance, it is approved for organic as well as conventional use, and controls many serious pests not otherwise easily controlled. He calls it a "public good," but suggests that engineering it into crops on a massive scale is the moral equivalent of loading everyone's toothpaste with antibiotics. Yes, the antibiotics would yield an immediate "benefit" in terms of reduced incidence of certain diseases. But the consequences for both immediate and long-term health would be ugly indeed, since disease microbes would develop resistance much more rapidly than otherwise. In the case of Bt, the inevitable development of resistance by pests will reduce the useful lifetime of this invaluable pesticide to a small fraction of what it would otherwise be. Then we'll be off to search for the next silver bullet.

It's a measure of the biotech industry's narrow and self-serving environmental assessment that the Bt toxin in the crop itself is never added into the calculations of pesticide use. Yet, speaking of Bt corn, Benbrook estimates that (depending on how you frame the question) there is 10 to 10,000 times as much Bt toxin produced in the crop as would have been applied in the usual external applications—and that's assuming a year in which the corn borer *needed* to be controlled at all.

Moreover, researchers have discovered that the Bt toxin released by the crop into the soil binds to soil particles and is then highly resistant to biodegradation. The implications for beneficial soil organisms are almost completely unknown—although the researchers found that a high percentage (90–95 percent) of insect larvae exposed to the toxin died (Saxena et al. 1999).

Crops genetically modified for resistance to herbicides pose similar problems. Knowing that their crops will more or less tolerate an herbicide, farmers are not likely to *reduce* their applications. Monsanto has requested and received from the Environmental Protection Agency a threefold increase in allowance for glyphosate residue on Roundup Ready soybeans. (Glyphosate is the active ingredient in the company's Roundup herbicide.) The increased residues are hardly an environmental improvement. While glyphosate is generally considered relatively harmless for humans and vertebrate animals (Williams et al. 2000), an increasing number of studies demonstrate harmful effects. Glyphosate has been linked to non-Hodgkin's lymphoma, a cancer of white blood cells (Hardell and Eriksson 1999), and has been shown to be toxic to animal cells (Marc et al. 2002, Peixoto 2005) and to human placental cells (Richard et al. 2005). The latter two studies point to an additional concern, which Peixoto (2005) describes: "Virtually, every pesticide product contains ingredients other than those identified as the 'active' ingredient(s), i.e. the one designed to provide the killing action. These ingredients are misleadingly called 'inert.' Commercial glyphosate formulations are more acutely toxic than glyphosate."

The vast expansion of acreage in herbicide-resistant crops has led to huge increases in the use of glyphosate—between 1996 and 2004 an increase of 138 million pounds (Benbrook 2004). This large-scale adoption of single-pronged weed-control strategies is deeply troubling because it encourages herbicide resistance in weeds (already observed with glyphosate) and wholesale shifts in weed populations. Plant scientist and expert on herbicide resistance Stephen Powles says, "There is going to be an epidemic of glyphosate-resistant weeds. In 3 to 4 years, it will be a major problem" (cited in Service 2007). Widespread glyphosate resistance will require new and additional herbicides, and the resulting treadmill, as Benbrook puts it, "is on hyperdrive today. We'll burn up the current generation of herbicides in five, ten, or fifteen years instead of three to five decades."

The alternative to the treadmill is to turn our attention away from quick fixes and look at ecological integrity. Mary-Howell Martens, who was formerly a genetic engineer and conventional farmer, now farms 1,100 acres organically in New York state. Like many other organic growers, she and her husband, Klaas, grow soybeans without using any herbicides. They work instead with nature, relying on soil fertility (the calcium-magnesium ratio in particular affects weed vigor); long, diverse rotations, including corn, soybeans, clover, and grains, to disrupt weed cycles; weed-free seed; well-timed tillage early on, so that the crop gets ahead of the weeds and tends to smother them; and avoidance of high-salt fertilizers, since salt compounds stimulate weed growth. Later weed control can be done mechanically, on a spot basis, as needed.

Orchestrating Nature's Complexity

Most people regard genetic engineering as the future of agriculture, if only because it is sophisticated, cutting-edge science. But impressive procedures in the laboratory do not automatically equate to precise effects upon nature. Even if it were true that DNA presents us with a kind of master computer program controlling the living organism, every software engineer knows about the unpredictable and sometimes disastrous consequences for massively intricate programs when someone goes in and "twiddles the bits." In 1976, when computer programs were vastly simpler than today, MIT computer scientist Joseph Weizenbaum could write a now-classic chapter entitled "Incomprehensible Programs," where he pointed out that any substantial modification of a large, complex program "is very likely to render the whole system inoperative" (Weizenbaum 1976).

In its application to agriculture, genetic engineering is crude, blindfolded, trial-and-error science—and not only because the consequences of particular genetic alterations are largely unknown. The farmer is often prevented from exercising skilled judgment based on the ecological realities of the local environment.

Take, for example, the farmer who plants Bt corn as protection against the European corn borer. He commits to round-the-clock, season-long application of a pesticide in his fields before he knows whether the corn borer will even be a problem. In major parts of the corn belt, the answer is that, during most seasons, it will not.

If you really want technical sophistication, don't look at the latest biotech application, but at the many successes of Integrated Pest Management (IPM). IPM is founded on decades of painstaking investigation into the incredibly complex and subtle weave of natural ecologies. Where the main trend of today's biotech agriculture is to isolate the farm from its environment, reducing the operation to the simplistic terms of a few manageable variables, IPM at its best tries to work *with* the environment, penetrating the boundless complexity with an understanding that can turn intricate equilibria to good use.

The task is not so easy. It is well worth noting the attempts at biological pest control that have gone awry. The authors of a review study ask whether "nontarget effects" might be "the Achilles' heel of biological control" (Lauda et al. 2003). For example, a Eurasian weevil was introduced in 1969 to North America to help control the spread of weedy thistles of European origin. It was known that this weevil lays its eggs in the flowerhead buds of thistles and that the larvae feed on the developing flowers, thereby preventing seed production. Oddly, no testing on native North American thistle species was carried out prior to introduction. By 2001, the weevil had spread to twenty-two of the over ninety North American species of *Cirsium* thistles. Populations of the indigenous Platte thistle in Nebraska declined dramatically, and a fruit fly that feeds on thistles also decreased in numbers.

This is exactly the kind of lesson we need to take to heart in the era of genetic engineering. Both the successes and failures of IPM drive home the importance of reckoning with an entire context instead of aiming at single cause-and-effect changes. The ecologically oriented integrated pest manager both recognizes and accepts this challenge, leading to a greater sense of caution and to a different way of working.

It is, after all, one thing to take the heavy-handed biotech approach and engineer a pesticide into every cell of a crop, and quite another to manage the ecological interrelationships of the farm so that the offending insect is controlled by the natural balances of the larger context. Tragically, the more simple-minded, heavy-fisted approach tends to destroy the possibilities inherent in the more subtle practice. Among other problems, converting an entire crop into a pesticide virtually guarantees the rapid emergence of pest resistance, which IPM has taken such pains to avoid.

Working with natural complexity rather than against it is the aim of a remarkable research organization in Kenya, the International Centre of Insect Physiology and Ecology (ICIPE). The Centre brings together molecular biologists, entomologists, behavioral scientists, and farmers in an interdisciplinary effort to control the various threats to African crops.

The most important pests of corn and sorghum on that continent are the stemborer and witchweed (*Striga*), which, together, can easily destroy an entire crop. ICIPE researchers developed a "push-pull" system (ICIPE 2002/2003; Khan et al. n.d.): a grass planted outside the cornfield attracts the stemborer, while a legume planted within the cornfield repels the insect and also suppresses witchweed by a factor of forty compared to a corn monocrop—all while adding nitrogen to the soil and preventing erosion—and, finally, an introduced parasite radically reduces the stemborer population. As if this were not enough, the grass and the legume can be fed as fodder to the farmers' livestock.

ICIPE director Hans Herren won the World Food Prize in 1995 after the Centre gained control over the mealy bug that threatened the cassava crop, a staple for 300 million people. (A small, parasitic wasp was instrumental in the success.) No chemical applications and no costs to the farmers were involved. Yet Herren doubts he could obtain funding for such a project now. "Today," he says, "all funds go into biotechnology and genetic engineering." Biological pest control "is not as spectacular, not as sexy" (quoted in Koechlin 2000b).

The Real Future of Agriculture

Fortunately, some work on Integrated Pest Management continues, and the results are often so dramatic that one wonders why the genetic engineering labs have secured all the glamour for themselves. Even the simplest step toward balance sometimes yields striking results. In what the *New York Times* called "a stunning new result" from a vast Chinese agricultural experiment, tens of thousands of rice farmers in Yunnan province "have doubled the yields of their most valuable crop and nearly eliminated its most devastating disease—without using chemical treatments or spending a single extra penny" (Yoon 2000).

The farmers, guided by an international team of scientists, merely interplanted two varieties of rice in their paddies, instead of relying on a single variety (Zhu et al. 2000; see also Zhu et al. 2003). This minimal step toward biodiversity led to a drastic reduction of rice blast, considered the most important disease of the world's most important staple. The fungicides previously used to fight rice blast were no longer needed after just two years.

The experiment, covering 100,000 acres, "is a calculated reversal of the extreme monoculture that is spreading throughout agriculture, pushed by new developments in plant genetics," observed Martin S. Wolfe in a commentary in *Nature* (Wolfe 2000). The problem, Wolfe suggests, is that monocultures provide a field of dreams for the development of super pests. The conventional solution—to breed resistant varieties and develop new fungicides—leads to rapid pest resistance. "Continual replacement of crops and fungicides is possible, but only at considerable cost to farmer, consumer, and environment."

These costs make the virtues of the new rice system all the more dramatic. How was rice blast overcome? Researchers, Wolfe says, have identified several factors in play. To begin with, a more disease-resistant crop, interplanted with a less resistant crop, can act as a physical barrier to the spread of disease spores. Second, when you have more than one crop variety, you also have a more balanced array of beneficial and potentially harmful pests that hold each other in check. A single pathogen, such as the one involved in rice blast, is therefore less likely to gain the upper hand.

Also, of the two varieties of rice used in the Chinese experiment, the taller variety was the one more susceptible to blast. But, when planted in alternating rows with the shorter variety, the taller rice enjoyed sunnier, warmer, and drier conditions, which appeared to inhibit the fungus.

And, finally, a kind of immunization occurs when crops are exposed to a diversity of pathogens. Upon being attacked by a less virulent pathogen, a plant's immune system is stimulated, so that it can then resist even a pathogen that it would "normally" (that is, in a monoculture) succumb to.

This last point reminds us that disease susceptibility is not a fixed trait of a crop variety, but relative to the conditions under which the crop is grown. Many existing susceptibilities reflect the crop's extreme isolation from anything like a natural or supportive environment, with its checks and balances. This environment includes not only other plants, but also the complex, teeming life of the soil—life that is badly compromised by "efficient" applications of fertilizers, herbicides, and pesticides. And, as these new findings indicate, even a "healthy" variety of disease organisms is important. What biotech company, focused on the latest, profit-promising lethal gene, would encourage such a balanced awareness among farmers?

Harnessing Complexity

When biotech proponents say, as they often do, "Prove to us that anyone has died or been made seriously sick by genetically engineered foods," the pathology is in the question itself. The underlying stance is, "If you can't show us the corpses, where the hell's the problem?" This suggests a complete avoidance of the ecological, social, economic, and ethical questions posed by the whole trend of technological agriculture.

If the right questions were being asked by those pushing biotech on farmers, they would be saying, "Look, here's why we think this kind of crop—and farm, and business structure, and community—is better for society than a highly diversified, local, small farm–based, organic agriculture."

But they do not address this larger picture, continually drawing our attention instead to particular technological achievements. They offer the farmer specific "solutions," but, as Amory Lovins, cofounder of the Rocky Mountain Institute, has remarked, "If you don't know how things are connected, then often the cause of problems is solutions" (Lovins

2001). Nor are they quick to mention the one way their systems *do* surpass all alternatives: they offer more patent opportunities for biotechnology concerns. It's hard to package all the local variations and contingencies of an environmentally healthy agriculture into a proprietary, uniform, for-all-purposes commercial system.

The question is why we would *want* such a package. The assembly-line uniformity and near-sterility of those endless Nebraskan cornfields certainly do appeal to some of our current inclinations, but they are not the inclinations of nature. It's true that we must work creatively upon nature. But eliciting the yet-unrealized potentials of an ecosystem is one thing; firing silver bullets at it is quite another. We have scarcely begun to understand all that nature can teach us about the bounty of the earth, and it would be a shame for the students to attempt an ambitious reengineering of the teacher before they have learned what she knows.

Chapter 2

Golden Genes and World Hunger

Let Them Eat Transgenic Rice?

Having become disenchanted with the early hype about genetic engineering, we were struck by the announcement in 1999 of a new genetically engineered crop that looked less like an arbitrary exercise in the manipulation of nature than an altruistic attempt to improve the human condition. If biotechnology can display beneficent potentials, how better to do it than by placing a daily bowl of genetically engineered "golden rice" on the dinner tables of millions of Asian children, thereby saving them from immense suffering?

This hope, many researchers believe, is now nearing fulfillment. But a full conversation around that envisioned bowl of rice has yet to occur. And until it does occur, we will have no means to assess the technical achievements represented by the bowl. In what follows we venture some preliminary contributions toward such a conversation.

Beyond Frankenfoods

Transgenic golden rice does not yet fill the bowls of hungry Asian children. But the possibility that it will has been the bright hope of scientists and biotech companies beaten down by the consumer backlash against the rapid and largely covert introduction of genetically modified organisms into global food supplies. The advertisement for golden rice, widely broadcast, is that it avoids all the pitfalls associated with the ill-fated "Frankenfoods" that so unsettled the buying public. What lends this new, experimental rice its golden color is the presence of beta-carotene within the part of the kernel—the endosperm—that remains behind (normally as "white rice") after milling and polishing (Ye et al. 2000; Paine et al. 2005). Beta-carotene is a precursor of vitamin A; the human body can use it to form the vitamin. This is important because millions of children, especially in Asia, suffer from vitamin A deficiency, which can lead to blindness.

By most accounts the virtues of golden rice are many:

- It is not the product of profit-seeking biotech companies. The original research, funded by the Rockefeller Foundation, the Swiss government, and the European Union, was performed at Swiss and German universities.
- Golden rice researchers stress that once the rice is available for field plantings, it will be freely distributed to poor farmers in the third world. It's a rather grotesque symptom of the biotech revolution that seventy different patents from thirty-two companies are "attached" to biotech golden rice. Agreements have been reached with these companies to forgo royalties so that seeds can be donated to farmers who do not earn more than $10,000 annually—which is the case with most third-world rice farmers. Those who earn more will have to buy the seeds and pay royalty fees.
- Rice naturally makes beta-carotene and other carotenoids, which are present throughout the plant—except in the endosperm. The genetic manipulation producing golden rice is simply designed to extend this natural production of beta-carotene into an additional part of the plant. In her commentary on this research in *Science*, Dartmouth biologist Mary Lou Guerinot suggests that the fears of most opponents of genetically modified foods will be allayed by the new rice (Guerinot 2000). After all, it's a far cry from transferring fish genes into plants.
- Unlike with many of the current genetically modified organisms, golden rice poses no risk of increased resistance to herbicides or insecticides.
- And, of course, the primary virtue of golden rice is its announced potential for solving problems of hunger and malnutrition in developing nations. Such a purpose hardly seems gratuitous or grasping. Who could possibly object?

So golden rice, as we now hear the story, looks rather like an almost magical solution to a major problem. Because golden rice seemed in fact to be a positive example of how to apply genetic engineering in agriculture, we looked into it more carefully. It turns out that the situation is much more complex than the usual story allows.

The immediate challenge for researchers is to develop hardy strains of the transgenic rice—the first field tests started in 2004 on the fields of Louisiana State University's AgCenter Rice Research Station. But this barely touches upon the conversational complexities the researchers must negotiate if they wish to enter constructively into the modern contexts of hunger and malnutrition. Here, briefly, are a few of the themes that need taking up.

If You Grow the Rice, Can You Deliver It to Those Who Need It?

The sobering fact is that "nearly eighty percent of all malnourished children in the developing world in the early 1990s lived in countries that boasted food surpluses" (Gardner and Halweil 2000, 17). The Green Revolution in Asia brought about a shift toward intensive cultivation of fewer crops, like wheat and rice, which are often grown for export. Traditional diverse polycultures have yielded to large monocultures.

At the same time—and at least in part due to the Green Revolution and other technology-driven changes—hundreds of millions of people have migrated from rural to urban areas in Asia during the past few decades. Mostly poverty-stricken, these transplants take up residence in the ever-expanding slums around cities and can't buy the food they need. Golden rice will do them no good if they can't afford it—and if they can afford it, then it is not clear what the new rice offers that would not be offered better by a more traditional and diverse diet.

Every green part of a plant contains beta-carotene. When Indian scientist and activist Vandana Shiva was asked at a lecture what alternative she saw to golden rice, she cited "the 200 kinds of greens we grow on our farms" (see also Shiva 2000). Traditional cultures never subsist on rice alone. In addition to the many different types of greens grown in India, wheat, millet, and various legumes are cultivated, not to mention the wild greens gathered from the countryside. Such polycultures develop differently in each region, but all allow, as long as there is enough food, for a balanced, life-sustaining diet.

It needs recognizing that what we in the western world embrace as export-driven economic growth has contributed to the problem of hunger in developing nations (Lappé et al. 1998, Rosset 2005). Golden rice can be seen in part as a one-dimensional attempt to "fix" a problem created by the Green Revolution—namely, the problem of diminished crop and dietary diversity. But the fix offers no direct help to those who have been displaced by the revolution and who cannot buy the food they need.

There are alternative approaches that do more justice to the complex geographical, historical, social, political, and economic issues. In 1993 the Food and Agriculture Organization of the United Nations, collaborating with nongovernmental organizations such as Helen Keller International, began a program to help poor people in Bangladesh grow a diverse array of plants to combat vitamin A deficiency (reported in Koechlin 2000a). In areas where people have at least small plots of land, families—usually mothers become the driving force of such projects—were introduced to different carotene-rich varieties of fruits and vegetables, and they learned cultivation methods. Landless families were shown how they could plant vines in pots on outside walls. They then planted beans and squashes that can grow up the vines.

When women noticed the positive health effects of their new diet, news spread by word of mouth, and now approximately 600,000 households (about 3 million people) participate in this project. This is, relatively speaking, a small number, but the project is promising because it can become part of cultural tradition. It empowers people instead of making them dependent on western aid. Scientists evaluating the project found that the general health of the participants improved and that even small plots can provide sufficient vitamin A in the diet. Moreover, the more different kinds of fruits and vegetables people ate, the better the uptake of carotene—an illustration of the inherent value of natural variety in the diet.

After assessing a number of such projects, John Lupien of the Food and Agriculture Organization concluded: "A single-nutrient approach toward a nutrition-related public health problem is usually, with the exception of perhaps iodine or selenium deficiencies, neither feasible nor desirable" (quoted in Koechlin 2000a).

If You Deliver the Rice, Will They Eat It?

"We must not think," writes Jacques Ellul, "that people who are the victims of famine will eat anything. Western people might, since they no longer have any beliefs or traditions or sense of the sacred. But not others. We have thus to destroy the whole social structure, for food is one of the structures of society" (Ellul 1990, 53).

Billions of Asians subsist on rice, which they mostly consume as white rice. To obtain white rice you must first remove the husks from rough or paddy rice, leaving the brown rice kernel. Then you must remove the embryo and bran layers by milling and polishing. These discarded, nutrient-rich layers happen to contain carotene. What is left after polishing is the shiny white endosperm—mainly starch.

This raises the obvious question: why not solve the problem of nutritionally inadequate rice by getting people to eat brown rice, containing protein, carotene, and various micronutrients?

The issues, again, are complex. Brown rice does not keep well in the humid South Asian climates, which is the reason scientists usually cite for Asians eating white rice. But while most rice is milled and sold as white rice, the rough rice kernel—still enveloped by its husk—can in fact be stored for long periods. The agronomist Heinz Bruecher observed that "the small farmer in Asia proceeds differently and avoids polishing by husking only as much rice as he needs at a time. In this way he always has a nutritious grain in storage" (Bruecher 1982, 58). Perhaps this practice could be encouraged.

But we must also reckon with the cultural traditions related to white rice. In Asia, rice is not just something that is ingested in the way we eat french fries. It is steeped in thousands of years of culture and tradition. Different shapes, sizes, and cooking consistencies are preferred, depending on the context: everyday rice, rice for special occasions, rice for flour, rice to accompany other specific foods, and rice for ceremonies. The whiteness of rice also has spiritual connotations:

> There is more to eating than merely ingesting nourishment to survive, more to living than merely surviving. Confucius in 500 BC knew this well as he preached the gospel of a virtuous, yet graceful life. He was a stickler for excellence and ceremony at the table and insisted on the pure whiteness of rice in sheer, elegant porcelain bowls as a background for light emerald-green vegetables picked at their succulent zenith, golden brown stir-fried morsels of duck, pork or fish, and deep red jujube dates. "Come eat rice with me" is the most gracious greeting in Chinese hospitality. In old China, families kept two crocks of rice,

> a large one of gleaming, white polished rice for the family, a smaller one of coarse brown rice for seeking one more day of existence. (Gin 1975)

The sensory symbolism of "pure whiteness" and "emerald-green" shows how a religious culture judges food as a spiritual-physical reality. The diet Confucius recommends is, in more prosaic terms, nutritionally balanced. People who use white rice experience it as being lighter and easier to digest, and find that it allows the taste of other foods to come to the fore. It is prepared in many different ways. In the context of a varied diet, white rice is an integral part of Asian cuisine.

Only the beggar receives the more nutritious brown rice—but without anything else—allowing him to eke out one more day. So it is that white rice can become a symbol for high social and economic status in Asian cultures. When the poor emulate the rich by consuming white rice, they are actually putting their already precarious health in greater danger. In this way social inequality accentuates nutritional problems.

It would be reasonable to encourage the use of brown rice throughout Asia, but any such program must reckon with deeply rooted cultural traditions. Certainly the new golden rice will bump up against these traditions, and it is not at all clear how the resulting conversation will play itself out. If we wish to engage in the conversation at all, the question is whether it makes more sense to push the one-dimensional "solution" offered by golden rice, or instead to cultivate the potentials of a traditional, diverse diet, possibly in conjunction with greater use of brown rice.

If They Eat the Rice, Will It Do Them Any Good?

If golden rice replaces white rice in the Asian diet, can we be sure this will solve the vitamin A deficiency problem? That is, leaving the social issues aside, will the proposed solution at least achieve its narrow aim?

Not necessarily. It is a naive understanding of nutrition—encouraged by a habit of input-output thinking—that says you can add a substance to food and the body will automatically use it. Beta-carotene is fat-soluble, and its uptake by the intestines depends upon fat or oil in the diet (Erdman et al. 1993). White rice itself does not provide the necessary fats and oils, and poor, malnourished people usually do not have ample supplies of fat-rich or oil-rich foods. If they were to eat golden rice without fats or oils, much of the beta-carotene would pass undigested through the intestinal tract.

Moreover, fats and also enzymes (which are proteins) enable carotene and vitamin A to move from the intestines to the liver, where they are stored. Proteins are bound to the vitamin in the liver, and enzymes are again required for transport to the different body tissues where the vitamin is utilized. A person who suffers protein-related malnutrition and lacks dietary fats and oils will have a disturbed vitamin A metabolism.

In sum, carotene uptake, vitamin A synthesis, and the distribution and utilization of vitamin A in the body all depend on what else a person eats, together with his physiological state. You can't just give people more carotene and expect results. There is no substitute for a healthily diverse diet.

Who Will Grow the Golden Rice?

Of the many thousands of rice varieties grown in Asia, most are local land races. Despite the introduction of high-yielding varieties in the Green Revolution, Indian farmers still use traditional varieties in over 58 percent of the rice acreage (Kshirsagar and Pandey 1997). These varieties serve their desire for different types of rice, while also providing the diversity needed within local ecological settings. The number of varieties a farmer grows tends to increase with the variability of conditions on the farm.

For example, when they don't irrigate, farmers in Cambodia plant varieties with regard to early, medium, and late flowering and harvesting dates; eating qualities (such as aroma, softness, expansion, and shape); potential yield; and cultural practices (Jackson 1995). In India a farmer might have high, medium, and low terraces for planting. The low terraces are wetter and prone to flooding; they are planted with local, long-growing varieties. In contrast, the upper terraces dry out more rapidly after the rains, so farmers plant them with drought-resistant, rapidly maturing varieties. Altogether a farmer may plant up to ten different rice varieties—a picture of diversity and dynamic relations within a local setting (Kshirsagar and Pandey 1997).

This multiformity has evolved locally and regionally over long periods. Since the Green Revolution, more and more farmers plant, in addition to land races, high-yielding varieties. The price they pay for this progress is dependence on irrigation, fertilizers, and herbicides. The use of insecticides has become widespread, although they have been shown to be ineffective (Pingali et al. 1997, chapter 11). (Sometimes the highest recommendation for western, industrial-style agricultural practices in the third world is that they are "modern.") The locally evolved land varieties, in contrast, tend to be more drought- and pest-resistant.

Imagine transgenic golden rice in this context. Agronomists are currently breeding transgenic varieties that they will test under field conditions. If these prove viable, large-scale seed production could begin and also interbreeding with other varieties. If bred into high-yielding varieties, golden rice would be grown primarily on large, export-oriented farms. In this case the rice would do little to alleviate Asia's food problems—and, who knows, it might even end up being exported to America and Europe. As biologist Margaret Smith points out, "If golden rice is to have the impact on vitamin A deficiency that its proponents claim, it will need to be very widely grown, just as the original 'green revolution' rice varieties [were] a few decades ago. We now have a sense of the value of that earlier loss of diversity, and should aim to avoid repeating that history" (Smith 2005).

One thing is clear: unless the development of such a new rice variety is embedded within the context of changing social and economic policies, history will repeat itself. And even if the golden rice DNA is introduced via breeding into varieties that small farmers use, these new, transgenic varieties will be subject to local practices and conditions. What started out as an isolated laboratory variety would gradually intermix and change, probably looking very different in different places. Whether the genetic alteration would prove stable in the midst of this flux is a real question. Although no one can say what will happen, one *can* say: things will change. It is unrealistic to think that genetically engineered plants are immune to context.

What Will Rice Make of Its Golden Genes?

The fundamental problem with genetic engineering from the very beginning has been the absence of anything like an ecological approach. Genes are not the unilateral "controllers" of the cell's "mechanisms." Rather, genes enter into a vast and as yet scarcely monitored conversation with each other and with all the other parts of the cell. Who it is that speaks through the whole of this conversation—what unity expresses itself through the entire organism—is a question the genetic engineers have not yet even raised, let alone begun to answer (see chapter 11).

But without an awareness of the organism as a whole, we can hardly guess the consequences of the most "innocent" genetic modification. The analogy with ecological studies is a close one. As we discussed in the previous chapter: change one element of the complex balance—in an ecological setting or within an organism—and you change the whole. It is a notorious truth that our initial expectations of an altered ecological setting often prove horribly off-target. And the possibility of improving our discernment depends directly upon our intimate familiarity with the setting as a whole in all its minutia and unity.

Certain herbicides kill plants by bleaching them—that is, by disrupting carotene metabolism and blocking photosynthesis. When scientists genetically altered tobacco plants to give them herbicide resistance, some of the plants indeed proved resistant to an array of herbicides (Misawa et al. 1994). Unexpectedly, however, leaves of the transgenic plants produced greater amounts of one group of carotenes and smaller amounts of another group, while the *overall* carotene production remained about normal. In some unknown way the genetic manipulation affected the balance of carotene metabolism, but the plant as a whole asserted its integrity by keeping the overall production of carotene constant.

Such unexpected effects are typical, expressing the active, adaptive nature of organisms. An organism is not a passive container we can fill up with biotech contrivances. Even when scientists try to change the narrowest trait of an organism, the organism itself responds and adapts as a whole.

Recently scientists have developed the technique of metabolic profiling to detect whether substances other than those targeted are changed by a genetic manipulation. In one experiment different varieties of genetically engineered potatoes were created that break down the sugar sucrose in different ways (Roessner et al. 2001). This entails a small genetic change that is associated with the production of a different specific enzyme in each of the transgenic lines. They then investigated the amounts of eighty-eight different substances (starch, different sugars, different amino acids, and so on) being produced in the tubers. Surprisingly, there was not just a change in amount of the substances in the specific breakdown pathway affected by the genetic manipulation, but in most of the eighty-eight substances. All the transgenic varieties differed from the nonmanipulated potatoes. For example, the transgenic potatoes often produced more amino acids than the nonmanipulated potatoes, and nine substances were found in the transgenic potatoes that could not be detected in the nonmanipulated potatoes. So what was intended to be a narrowly circumscribed alteration in a metabolic pathway had ripple effects on the entire metabolism in ways no one would expect.

The transgenic golden rice plants were reported to be "phenotypically normal" (Ye et al. 2000). This statement needs to be read: "no visible modifications were noted." The researchers didn't undertake a biochemical analysis of the kernels to see how their overall content might have changed. What *doesn't* a golden rice kernel produce as a result of the plant's breaking down excessive amounts of carotene? What new substances does it produce? And what are the changed balances among substances normally present? The more one learns about the flexible and dynamic nature of organisms—demonstrated so clearly by genetic engineering experiments themselves—the more one comes to expect the unexpected and to realize that we cannot know what subtle effects a manipulation may have.

How many genetic engineers have pondered the remarkable fact that rice, despite the myriad varieties that have arisen over thousands of years, never produces carotene in the endosperm of the kernel? The rest of the above-ground plant makes carotene, and the endosperm should (according to prevailing conceptions) have the genes that would allow it to produce carotene. But it never does so. Certainly that should give us pause to consider what we're doing. Might the excess carotene in the seed affect in some way the nourishment and growth of a germinating rice plant? What does it mean to force upon the plant a characteristic it consistently avoids? Can we claim to be acting responsibly when we overpower the plant, coercing a performance from it before we understand the reasons for its natural reticence?

Organisms are not mechanisms that can be altered in a clear-cut, determinate manner. The fact is that we simply don't know what we're doing when we manipulate them as if they were such mechanisms. The golden kernels of rice almost certainly herald much more than a novel supply of beta-carotene.

A Disproportionate Interest in One-Shot Fixes

We often hear that biotechnology is merely doing what high-yield breeding, industrial agriculture, and nutritional science have done all along—but now much more efficiently. In one sense that's exactly right and also exactly the problem: we don't need more of the same. What we need is to overcome an epidemic of abstract, technological thought that conceives solutions in the absence of organic contexts. We need a refined ability to enhance life's variety rather than destroy it. And we need to realize that the problems of life and society are not malfunctions to be fixed; they are conversations to be entered into more or less deeply. The more deeply we participate in the conversation, the more thickly textured and revelatory it becomes, reacting upon all the meanings we bring to the exchange.

The engineering mind-set that tries to insert individual traits into rice by manipulating particular genes is closely allied to the long-standing agricultural mind-set that tries to improve crop yields in a purely quantitative sense by injecting the right amounts of NPK (nitrogen, phosphate, and potash) into the soil. On this view the soil offers little more than a structural support for the roots. At the same time, it is a kind of hydroponic medium into which we place the various "inputs" that we can identify as requirements for plant growth. What this approach overlooks is . . . well, just about everything. Fixated upon inputs, outputs, and uptake mechanisms, it loses sight of the unsurveyed, nearly infinite complexity of life in a healthy, compost-enriched soil. The truth of the matter is that whatever we can do to enhance the diverse, living processes of the soil will likely improve the quality of the crop, and yet an input-output mentality proceeds to destroy the life of the soil through simple-minded chemical applications. Our one-shot solutions, much too narrowly targeted, rip through the fabric of the life-sustaining context.

Sponsors of the green and genetic revolutions are not inclined to ask what is lost when input-intensive, high-yield monocultures replace the kind of local diversity that resulted in thousands of local rice varieties throughout Asia. We have never heard a biotechnologist venture the thought that local varieties may actually—through their long history of co-evolution with the people who bred them—be uniquely adapted to the nutritional needs and dietary complexities of the local population.

The adaptation is not hard to imagine when you consider beta-carotene. Plants make many different types of carotene; beta-carotene is only one member of a large family of substances. Each species of green, squash, or brown rice produces its own unique array of carotenes, with different types and amounts arising in different tissues depending on changing conditions. Numerous species-specific carotenes have scarcely been investigated.

Similarly, human beings need different kinds of carotenes, and, as long as a reasonable diversity of crops is available, each individual will draw out of his food what he needs. But what if, in the name of this or that specific "input" abstracted from the complex, nutritional matrix of life, we proceed to destroy the matrix? The disproportionate hope placed in golden rice, together with its salesmen's casual disregard of biological and social context, suggests the likelihood of precisely such destructive consequences.

There are no single-shot fixes in any profound conversation, including the one around that bowl of rice. There is only a progressive deepening of meaning. If we prefer the satisfaction of unambiguous bits of information, then—whether we conceive those bits as genes or NPK or the dietary inputs of Asian children—we abandon the wholeness and coherence of the conversation altogether. We can, in this case, certainly proceed with our narrow programs of manipulation and control, which are what we have left when we give up on conversation. But the results will be no more satisfying than a diet of rice alone.

Chapter 3

Will Biotech Feed the World?

The Broader Context

When giving presentations about genetic engineering and agriculture, we find that one of the most frequent questions is about feeding the world. How are we going to feed a growing human population, when already many millions of people around the globe are undernourished and suffering from hunger and even starvation? On a planet of more than six billion people, more than 800 million are undernourished (Food and Agriculture Organization of the United Nations 2005). Proponents of modern industrial agriculture believe genetically engineered crops hold the promise of a new Green Revolution, a revolution that will bring higher yields and nutritionally enhanced crops to third world countries.

In 2004, the Food and Agriculture Organization of the United Nations (FAO) issued a report describing how biotechnology can "help significantly in meeting the food and livelihood needs of a growing population" (Food and Agriculture Organization of the United Nations 2004, vii). Since the FAO is known for its multifaceted efforts to empower small, poor farmers in the third world, this endorsement of agricultural biotechnology, which is currently driven by a few giant multinational companies, came as a surprise to many. It also generated a wave of opposition. An open letter to the FAO's director, Jacques Diouf, which was signed by many third world farmers and civil society organizations, derides the report as highly biased and as fodder for the biotech industry's PR machine (http://www.grain.org/front/?id=24). Clearly, the place of genetic engineering in efforts to feed the world is a hot topic, and the debate is highly polarized.

This chapter places the issue of feeding the world into a broader context to help give clearer perspectives on a very complex topic. If we concentrate only on a technological application and its promises and pitfalls, we lose sight of the real problems. To dispense with some illusions at the outset, let's begin by looking close to home.

Hunger in the United States

According to a U.S. Department of Agriculture study, during 2004 13.5 million American households (35 million people) did not always have an adequate supply of food (Nord et al. 2005). In 4.4 million of these households, the situation was bad enough for the study to speak of "food insecurity with hunger."

These are astoundingly high numbers for the largest food-producing country on the planet. In 2003, the United States exported 93 million metric tons of wheat, corn, and soybeans. Evidently, the copious amount of food produced had very little effect on whether people went hungry. Seventy percent of the grain harvested in the United States is fed to cattle, pigs, and poultry.

In the United States—as elsewhere—hunger and food insecurity are related to a lack of money to buy food. More than half of the 13.5 million food-insecure American households receive some form of assistance through food stamps, free school lunches, and food pantries. Without this (albeit inadequate) safety net, which is funded largely by the federal government, the extent of hunger in the United States would be much greater.

As one might expect, the most needy people are those with incomes below the poverty line (in 2007 set at $20,650 per year for a family of four), households with children (especially single-parent households), and minorities (African Americans and Hispanics). The problem of hunger in the United States is an extremely complex issue of poverty, discrimination, and social and economic policies and practices.

The boom in cultivation of biotech crops in the United States since the late 1990s (in 2005 over 120 million acres, which made up two-thirds of the soybean crop and about one-third of the corn crop) has done absolutely nothing to address these issues. Since 1999 there has been a yearly rise in the number of food-insecure households; between 1999 and 2004 the number of families without enough food increased by 2.5 million.

Of course, those who believe biotech will feed the world never take the United States as their example. They look to developing countries, where there are manifest food shortages and a lack of adequate agricultural infrastructure. So, it would seem, citing the United States is irrelevant. It is cited here, however, to make one point loud and clear: even if the wishful thought that biotechnology could increase food production in developing countries became reality, this is not the same thing as providing people with food. There remain the underlying issues of poverty, food distribution, and economic and social policies. This is where looking at a rich country like the United States, where millions of people are not adequately fed, is instructive. It dispels the illusion that producing more food alone will feed more people.

The Green Revolution and Industrial Agriculture in a Larger Context

The Green Revolution, beginning in the late 1960s, dramatically increased grain yields. Crops such as wheat and rice were bred to have short stems and produce more grains. As the father of the Green Revolution, agronomist Norman Borlaug, states, "the results speak for themselves": in 1965 wheat yields in India were 12.3 million tons; in 2000 the yield was a record 73.5 million tons (*Wall Street Journal,* December 6, 2000). This is a remarkable increase.

But the Green Revolution also had hidden costs. As Gordon Conway—former head of the Rockefeller Foundation, which provided substantial support to the Green Revolution—remarks: "Green Revolution technology allows plants to channel more photosynthate into grain production, dramatically increasing yields if fertilizer and irrigation are provided. But it diminishes other useful traits, such as vigorous deep roots, sturdy stems, and ability to compete with weeds. Asking African farmers to invest in Green Revolution technology meant asking them to invest in fragile plants in a harsh landscape. Cereal yields in Africa have barely increased over the past 30 years and stand at a meager 1 ton per hectare; per capita food production is stagnant" (Conway and Sechler 2000, 1685).

High-yielding Green Revolution varieties have high yields only under certain conditions. They require large doses of fertilizer and irrigation. Breeding for high yield brings the loss of other vital characteristics (such as the ability to compete with weeds). Moreover, when the crops are planted in large monocultures, there is no getting around the increased use of pesticides (herbicides, insecticides, and fungicides). In other words, Green Revolution agriculture means importing a whole environment that makes higher yields possible.

While per capita food production in South America and India, for example, has increased during the last three decades, the number of hungry people has increased at an even greater rate (Rosset and Mittal 2000 and 2001; see also Rosset 2005). This is one of the more grotesque "side-effects" of the Green Revolution. "India is faced with an unmanageable food glut. From a food grain surplus of 10 million tons in 1999, the stocks have multiplied to 42 million tons. Instead of distributing the surplus among those who desperately need it, the government either wants to find an export market or release it in the open market" (Rosset and Mittal 2000). Green Revolution crops have been planted in connection with policies aimed at increasing food export as a way of increasing national income. The problem is that such policies bring little to the poor and hungry. As a 1997 study by the American Association for the Advancement of Science found, 78 percent of all malnourished children live in countries that export food (cited in Lappé et al. 1998; see also Altieri 2000 and Rosset 2005).

In contrast to its recent report on food and biotechnology, a 2002 report by the FAO states that "world agricultural production can grow in line with demand, provided that the necessary national and international policies to promote agriculture are put in place. . . . Agricultural production could probably meet expected demand over the period to 2030 even without major advances in modern biotechnology" (Food and Agriculture Organization of the United Nations 2002).

By focusing most attention on higher yields, proponents of industrial agriculture not only give undue weight to one component of a complex issue, but also ignore the problems that come with the higher yields themselves. These problems become evident when we look at the Green Revolution's broader implications, which include:

- crop varieties that are more environmentally sensitive, that is, less well-adapted to local conditions
- dependence on high-energy inputs
- support of large farm operations at the cost of small, low-input farming
- export-oriented production
- increase in pesticide-related health problems
- greater water pollution (fertilizer and pesticide runoff)

Hidden Costs

In the first study of its kind, Jules Pretty, the director of the Centre for Environment and Society at the University of Essex in the United Kingdom, tabulated the costs of industrial agriculture that go beyond direct expenses such as buying seeds or machines (Pretty et al. 2000). He was interested in the so-called externalities. As he states, "an externality is any action that affects the welfare or opportunities available to an individual or group without direct payment of compensation" (Pretty 2001, 114). The external costs include the money needed to treat or abate problems such as pesticides in drinking water, greenhouse gases produced by agriculture (methane, carbon dioxide, and nitrous oxide), bacterial disease outbreaks from agriculture, BSE ("mad cow disease"), destruction of biodiversity, and so on.

Pretty and his research team spent a number of years estimating the costs of negative externalities that arise from agriculture in the United Kingdom. The total costs were more than £2.3 billion per year (about £200 per hectare farmed). This enormous figure is about the same amount as the overall net farm income for the United Kingdom in 1996, and smaller than the £3 billion in subsidies that the U.K. government gives to support agriculture. Pretty emphasizes that the cost estimates are conservative, since they often only include treatment of a problem and not the costs of its eradication. As he states, we pay three times for our food: we pay for it at the market, we pay for it in our taxes that go to subsidize farming, and we pay again to clean up the mess.

Pretty's study makes numerically visible the hidden costs of an unsustainable approach to agriculture. It is certainly naive to believe that this approach promises a long-term solution to world hunger.

Because industrial agriculture is basically a whole package of intended practices and unintended consequences, when it is imported into an existing agrarian culture its effects can be very destructive. Ecologist Carl Jordan describes an example of this in the dry Sahel region of Africa (Jordan 2002). The Marka are an ethnic group that has cultivated rice since prehistoric times. They plant native rice and have developed different varieties that they use at different times and for different soils. The knowledge about rice cultivation is held secret, "a hierarchical system prioritizes access to land, and the rules regulating access to common property have been encoded into local Islamic law" (527). In this way the cultivation of rice is woven into a whole ecological, historical, and social fabric. If you isolate rice production from this fabric, the whole fabric begins to dissolve.

This has occurred in some areas of the Sahel region, where development projects aim to increase rice production as a means of contributing to the growth of the national market economy. In other words, a new form of rice farming was implemented based on a Western economic model. Local rice varieties were supplanted by an Asian rice variety. The knowledge of its cultivation was held by "outsiders" and no longer by the indigenous culture itself. The "unscientific" approach of the Marka people was no longer needed. Land allocation changed to fit the agro-economic model.

In Senegal this kind of "development" led to the degradation of 25,000 hectares of rice farmland due to poorly constructed irrigation systems. As Jordan summarizes, "the transition to a market economy ignores the nature of Sahelian climate and soils and deprives traditional Marka groups of their ability to respond flexibly in times of environmental stress" (527).

This example isn't intended to hearken back to the "good old days" and to flatly reject any modern approaches. Rather, the question is: if something from the outside is brought into a culture, can it stimulate further evolution of indigenous practices rather than destroy those practices by replacing them with unsustainable "solutions" along with all their externalities? The central problem is that when something comes from the outside and replaces an indigenous practice, it tends by its very nature to spread as a kind of foreign body and ramify destructively into the environmental, social, and economic structures and processes of that land.

Biotech Agriculture: A Model for the Third World?

Large-scale commercial farming of genetically modified (GM) crops began in 1996 and has grown steadily since. In 2006, 252 million acres of cropland were planted with GM soybeans, corn, cotton, and canola that are herbicide tolerant (Roundup Ready crops) or produce an insecticide (Bt crops) or both. The vast majority (91 percent) of the GM crops are grown on large industrial farms in the United States, Argentina, Brazil, and Canada.

Since the whole process of developing a genetically engineered crop is very expensive, genetically modified seed is sold by biotech companies at a premium price to farmers, who pay around $15 to $20 per acre as a "technology fee." Farmers sign a contract in which they agree not to use the seeds produced by the GM crops; instead they buy new seeds and pay fees year after year. The U.S. Department of Agriculture and the Monsanto Company hold patents on a genetic technology that would make second-generation seeds sterile in order to guarantee that farmers do not save and replant seeds from their GM crops. The patents for this "terminator gene," as it became known, were awarded in the 1990s and aroused hefty protest. As of 2007, no GM plants containing the terminator gene are on the market.

The largest-selling biotech seed today is herbicide-resistant soybean. These plants have been manipulated to withstand spraying with the herbicide glyphosate—which the farmer buys from the same company that sells the seeds. So the farmers are doubly dependent on the companies. Between 1996 and 2004, glyphosate sprayed on fields with herbicide-resistant soybeans, cotton, and canola increased the overall use of this herbicide on these three crops by 5 percent (138 million pounds) (Benbrook 2004).

The heavy reliance on one herbicide has led to the development of an increasing number of herbicide-tolerant weeds that are no longer killed by glyphosate (see http://www.weedscience.org/in.asp, and also Service 2007). GM soybean farmers can only hope that biotech companies have a new line of herbicides and herbicide-resistant crops in development to cope (temporarily) with the new generation of weeds their previous practices unintentionally brought forth. This is clearly not a sustainable practice—unless we call sustainable the never-ending task of finding temporary fixes for problems we cause through our unecological use of technology.

Surprisingly, a 2002 study by U.S. Department of Agriculture economists found that U.S. farmers rapidly adopted herbicide-resistant soybeans "even though we could not find positive financial impacts in either field-level nor the whole-farm analysis" (Fernandez-Cornejo and McBride 2002, 24). Agricultural economist Michael Duffy reached similar conclusions in a study comparing the yields and costs of GM crops compared to conventional crops in Iowa (Duffy 2001). Herbicide-resistant soybean fields do not necessarily increase yields (Benbrook 2002, Bohner 2003). So the most widely used GM crop is not necessarily benefitting—in a narrow economic sense—the farmers who are using it.

It may be that the desire for a spotless, weed-free field and for the ease of applying only one herbicide outweighs the lack of economic gain. We should not underestimate, in addition, the pull of "progress"—American farmers are strongly invested in the industrial model of agriculture, and biotech crops are viewed as the newest tool for advancement. In this connection we could speak of an ideological dependence.

These examples illustrate how the dominant present-day application of genetic engineering in agriculture is essentially industry driven, increasing a farmer's dependence on biotech companies and the industrial model of agriculture. It is hardly a model for the millions of farmers who grow their crops on small farms and produce the bulk of the food for the rural poor of third world countries.

But just because GM agriculture today is largely industrial in scale doesn't mean it has to be so. This is what proponents of GM crops for the third world often point out. As Ismail Serageldin of the World Bank states, "Biotechnology can contribute to future food security if it benefits small-farm agriculture" (1999).

Clearly, herbicide-tolerant GM crops that increase dependence on herbicides would do little to empower independent small farmers. With insecticide-producing crops it appears, at least at first glance, to be another matter. Here farmers can actually reduce the amount of insecticides they spray, since the plant itself has become an insecticide. In its 2004 report, the UN's Food and Agriculture Organization gives some examples of how small farmers have profited from growing insecticide-producing GM cotton as a cash crop.

The long-term sustainability of these insecticide-producing GM crops is highly questionable. That no widespread insect resistance to the plant-produced insecticide has yet arisen is largely due to the fact that farmers plant non-GM crops around the GM fields as a "refuge" for insects (Tabashnik et al. 2005). However, a new resistant strain of cotton bollworm was reported in 2005 in Australia, where GM cotton has been planted since 1996 (Gunning et al. 2005). It is probably only a matter of time until resistance to the GM plants appears in different parts of the world—and then what? There may also be other ecological and agronomic negative effects of GM cotton, as some evidence in China suggests (Dayuan 2002).

When proponents of GM agriculture in the third world are confronted with the problematic nature of this first generation of GM crops, they usually refer to a new generation of crops illustrating the yet untapped potential of this technology. We are to imagine GM crops that are drought-resistant or tolerant to salt buildup in the soil. Since drought and salt buildup are major factors limiting cultivation in arid regions of Africa, where there is widespread hunger, wouldn't they be just what is needed? Or what about GM plants that contain essential micronutrients, such as iron or vitamin A or even vaccines against diseases such as hepatitis B or cholera? These are not pie-in-the-sky ideas. Each one has been implemented in some university or biotech lab around the world.

In the previous chapter we examined one example in detail—beta-carotene-enhanced "golden" rice. What we learned there is that even the most altruistic ideas and goals become problematic when they are not considered within a broader social, economic, ecological, and organismic context. The intention to effect discrete, single-target changes in an organism lies at the heart of genetic engineering. But this frame of mind, which assumes one-directional cause-and-effect mechanisms, is inherently unecological, since all biological and ecological processes involve reciprocal relations. The key question is: Is it possible to overcome the limitations inherent in a technological approach to finding "solutions" to problems? Clearly, if the "problem" is not one to be solved with some new "solution," then the task is to stop thinking in these terms. The fascination with high-tech approaches only diverts our attention from more contextual ways of looking at the issue of world hunger.

Addressing Hunger in the World—Entangled Relations and Synergy

In its 2005 report *The State of Food Insecurity in the World,* the FAO delineates eight "millennium development goals" that reflect some of the challenges in third world countries (Food and Agriculture Organization of the United Nations 2005). The goals are to:

- eradicate extreme poverty and hunger
- achieve universal primary education
- promote gender equality and empower women
- reduce child mortality
- improve maternal health
- combat HIV/AIDS, malaria and other diseases
- ensure environmental sustainability
- develop a global partnership for development

Reducing hunger is seen as "an essential condition for achieving the other goals" and therefore is considered to be the central problem. But the report itself shows clearly that, just as having more to eat can improve, say, education and health, so also improving education and health care can reduce hunger. For example, during the past fifty years the state of Kerala in India has had a strong commitment to the education of women. Ninety percent of women are literate, and girls attend school at least until the age of fourteen. Although Kerala is not one of India's wealthier states (measured in terms of per capita gross domestic product), "it stands head and shoulders above the rest in terms of maternal and child nutrition and health." The report concludes that "promoting gender equality and empowering women could do more to reduce hunger and malnutrition than any other of the millennium development goals" (17).

The report emphasizes the need to increase agricultural production in rural areas of the third world, where three-quarters of the world's hungry live. But the focus is not primarily on increasing yields per se but on the larger context that affects agricultural production. One important area is governance: "internal peace, rule of law, rural infrastructure, and agricultural research are essential for increasing agricultural production and reducing hunger and poverty" (10).

What becomes crystal clear is that an avenue of approach may take as its starting point a particular "problem," such as combating disease. But to be fruitful in the sense that it contributes to weaving the fabric of a more sustainable culture, it needs to take into account how it is related to a larger whole, just as an organ is embedded in the dynamics of the whole body.

This is illustrated by a project called "Integrated Management of Child Health" (IMCA) that began in 1997 in a number of African countries. Previously, when a mother brought her child to a rural clinic, she met the "old 'factory-line' method where practitioners often made a quick guess at what was wrong with the patient and dispensed standard medication" (IDRC/CRDI 2004). The health care staff of these clinics is now trained to take more time to look at the whole child, including its diet and eating habits. Where a diet is lacking in micronutrients, such as vitamin A or iron, staff give supplements but also make suggestions concerning other foods the family members can themselves grow and eat. The program involves "a participatory process with the community [that] developed actions tailored to regional variations rather than a predesigned uniform strategy." There were "substantial improvements in health and micronutrient status in each of the five African countries including reduction in iron-deficiency anemia, sustained broad coverage with vitamin A supplements, improved dietary diversity and community development and empowerment."

Social ecologists Jules Pretty and Rachel Hine (2001) studied 208 projects in 52 countries in Africa, Asia, and Latin America that were using some form of sustainable agriculture practices. The study encompassed 8.98 million farmers working 28.92 million hectares (71.4 million acres) of land. Most of the farmers had small farms, with a typical household farming about 1.5 hectares (3.7 acres) of land.

Ninety-six of the projects had reliable information on food production, which could be compared with yields before sustainable practices were implemented. Small cereal farmers (rice, millet, sorghum, etc.) saw a rise in production from 2.33 to 4.04 metric tons per household per year. Small root crop (potato, sweet potato, and cassava) farmers saw their production more than double, from 11.02 to 27.5 metric tons per household per year. These remarkable figures show that relatively small changes in farming practices—such as using integrated pest management or improving soil fertility through composts—can lead to a rise in productivity.

But as Pretty and Hine discovered, most of the farmers intended much more than improving yields. The researchers describe the goals of sustainable agriculture in the following way:

> A more sustainable farming seeks to make the best use of nature's goods and services whilst not damaging the environment. It does this by integrating nature and regenerative processes, such as nutrient cycling, nitrogen fixation, soil regeneration and natural enemies of pests, into food production processes. It also minimizes the use of non-renewable inputs (pesticides and fertilizers) that damage the environment or harm the health of farmers and consumers. It makes better use of the knowledge and skills of farmers, so improving their self-reliance. And it seeks to make productive use of . . . people's capacities to work together to solve common management problems, such as pest, watershed, irrigation, forest and credit management. . . . Sustainable agriculture technologies and practices must be locally-adapted. They emerge from new . . . relations of trust embodied

> in new social organizations . . . , leadership, ingenuity, management skills and knowledge, capacity to experiment and . . . [innovation] in the face of uncertainty. (Pretty and Hine 2001, 37–38)

As Pretty and Hine point out, "each type of improvement, by itself, can make a positive contribution. But the real dividend is likely to come with appropriate combinations" (48). What's important is that by orchestrating the whole system, from soil fertility to credit financing, synergistic effects arise that make the whole more productive and stable.

For example, in a three-year project in the Jiangsu Province in China, rice farmers were supported in their efforts to transition from rice monocultures to rice-fish farming (Kangmin 1998). Growing fish in flooded rice paddies is an old practice that nearly died out in Southeast Asia. This was at least in part due to the increased used of short-stemmed, high-yielding "Green Revolution" rice varieties and the concomitant increase in the use of pesticides and fertilizers. In these fields there is too little water and too much poison for fish to thrive.

Li Kangmin, a scientist who assisted the farmers, describes the benefits of rice-fish farming:

> A rice field is a small artificial open ecosystem. The interaction between rice and fish has been called "waste not, want not," which indicates Chinese philosophy: The by-products or waste from one resource use must, wherever possible, become input into another resource use—an ecological principle. Culturing aquatic animals in rice fields can reduce the loss of nutrients in fields. Fish and other animals will help control pests and will loosen the soil as a result of their swimming and food searching activities, thus, aerating the soil, enhancing the decomposition of organic matter and promoting the release of nutrients from the soil. The excreta of aquatic animals directly fertilize the water in rice fields. (Kangmin 1998, 10)

During the three-year project, rice-fish farming was developed on 69,000 hectares (170,000 acres) of land. By the end of the three-year period the profit per hectare had increased 2.86 times compared to the previous rice monocultures. Not only did farmers have fish to consume and sell at local markets, but rice yields also grew by 10 to 15 percent. A welcome side benefit of the new practice was that the incidence of malaria dropped, since the fish were also feeding on mosquitos and their larvae.

This example illustrates that when we work with more complexity, interactions and "unintended consequences" arise that tend to have positive overall effects. In contrast, when we strive to decrease complexity—as in a monoculture where soil fertility practices are replaced with fertilizers—we create a system that tends to create more one-sided negative unintended effects, such as pollution and disease susceptibility.

Sustainable agriculture is much more than a set of new techniques. It provides the basis for a rejuvenation of land-based communities and greater food production and security. As Michael Stocking, an expert in tropical agricultural development, writes, "Interventions that use community-based approaches that empower farmers to manage their own situation therefore hold the greatest promise for maintaining soil quality and ensuring food security" (Stocking 2003).

Shifting Our Perspective

Feeding the world is not just a question of increasing yields. When we believe it is, we divert our attention from the much broader social, political, economic, and ecological issues influencing food production and hunger. If we continue to live under the illusion that we will find a technological solution to world hunger, and if we set our hopes on such solutions and channel our money and energy into their development, we can be pretty sure that world hunger will only grow.

What is needed is a shift in our way of viewing that can inspire and inform a different kind of practice. The shift means no longer thinking of the world's problems in terms of individual causes that can be manipulated or alleviated by single-target solutions. In the mode of thought that leads to industrial agriculture and genetic engineering we isolate "causes" out of a whole ecology and try to effect changes by manipulating these causes.

An ecological view takes a different approach. The focus is not on individual causes but on orchestrating the whole system. The whole is embodied in its interactions and in the synergies that arise out of these interactions. We attend to the reciprocal relations within the context of the whole rather than isolating linear pathways and manipulating them as though the rest of the system didn't exist.

In this way of viewing, there could be nothing more absurd—even if, on the short view, it produces lots of food—than, say, a factory farm housing thousands of pigs. The animals suffer under the narrowly confined conditions, unable to carry out many of their natural behaviors, such as rooting. The concentration of animals fosters health problems, leading to the widespread use of antibiotics. The pigs produce enormous amounts of feces and urine that are considered waste. "Stored" in lagoons, the fumes from this waste are a health hazard for people in neighboring towns, and the sewage pollutes ground and drinking water. Since huge amounts of corn-based feed are required, the corn, perhaps genetically modified, is grown in some other part of the country, and is fertilized with chemical fertilizers that also pollute streams and groundwater. It's hard to imagine a more unecological, unsustainable system. This is not a way to feed the hungry; it's a way to destroy the planet.

There is no fixed model for sustainable approaches. Since they are ecologically oriented, they must be adapted to local conditions, and those local conditions include the culture and people who live there. There can be no one grand "feed the world" plan. Ecological approaches to agriculture will take on different forms and different dimensions in different places; the many examples described in the Pretty and Hine study show this. As in other aspects of life, we learn through examples, which we modify to meet our situation, and then develop further.

What will be essential is that these kinds of approaches find the necessary economic and political support—that they at least be given the room to be tested and developed. When government farm subsidies are tied to acreage and commodity crops, as is the case in the United States, they directly support industrial agriculture at the cost of ecological farming. Much hinders the further, widespread development of an ecologically based agriculture. But there is no reason to doubt that true food security and the ecological health of the planet will depend on its taking root around the globe.

Chapter 4

We Label Orange Juice, Why Not Genetically Modified Food?

It is reasonable to expect that a label will tell you something significant about the food you buy. Because of widespread deceptive labeling, Congress began passing laws in the early 1900s to regulate food labels. The Food, Drug, and Cosmetic Act was first passed in 1938 and has been amended numerous times. In connection with the identity of food it states (in Section 401) that the FDA Secretary should make regulations when "such action will promote honesty and fair dealing with consumers." On its Web site the FDA declares that it is "one of the nation's oldest and most respected consumer protection agencies and describes its mission, which includes ensuring that foods are "safe," "wholesome," and "sanitary," and that food products also have "truthful and informative labels" (FDA 2004). The FDA has pages and pages of documents defining what should be on labels and what specific terms mean and don't mean (Code of Federal Regulations, n.d.).

A trip to the food store illustrates the FDA's influence. All products must clearly show the identity of their ingredients. The label must also identify any substances—such as preservatives or taste enhancers—that have been added to the food. The FDA has a database of more than three thousand food additives that must be named on labels. If there are concerns about the safety of the additive, such as in the case of saccharin, the label must contain a health warning for that substance.

The labels also make distinctions that producers might like to hide. Grape juice, for example, is 100 percent juice only if the label states "juice" without any qualifiers, such as "juice beverage" or "juice drink."

The latter terms indicate that the juice has been diluted or that flavorings have been added. And when the ice cream label states "vanilla flavored," you know that it contains an artificial substitute rather than natural vanilla.

Food labels also tell you something about the way a food has been processed. If the pasta sauce you buy has been heated (pasteurized) so that it will keep longer, it cannot be labeled "fresh," since the label "fresh" indicates that a food has not been processed. Similarly, when you buy orange juice, a label tells you whether the juice has been reconstituted by adding water to a concentrate. It states "from concentrate" to distinguish it from fresh-squeezed juice.

Another example of processing is radiation treatment, which is used to kill bacteria. Irradiated fruits and vegetables must carry the radura symbol on a label stating "treated with radiation." The FDA has "found it necessary to inform the consumer that irradiated food has been processed, because irradiation, like other forms of processing, can affect the characteristics of food" (Pauli 1999). Strangely, if an irradiated fruit or vegetable is used in a canned or packaged product, it need not be labeled, since the FDA reasons that consumers know they are buying a processed food. Evidently the FDA does not consider it important in this case to inform us about different kinds of processing.

Despite such anomalies, labeling rules show a good deal of common sense on the whole. The purpose of the label is to accurately inform consumers so that they know what they are getting and can make informed choices about the food they buy. The label should embody the intent of open and truthful disclosure. Insofar as this goal is achieved, the FDA is exercising its function as a consumer protection agency.

Genetically Modified Food: No Labeling

The FDA has declared that genetically modified foods or food ingredients need not be labeled. Why not? The FDA holds that genetically modified foods (GM foods) are essentially the same as traditional foods—there is "substantial equivalence," in the language of food scientists. For example, Monsanto scientists performed detailed nutritional analyses of traditionally bred soybeans and of "Roundup Ready" soybeans, which were genetically modified to be resistant to Monsanto's "Roundup" herbicide. The scientists compared nutrient content (protein, fat, carbohydrates, fiber, etc.), even analyzing the amounts of the different kinds of amino acids that make up proteins. They also compared so-called antinutrients (such as lectins), undesirable substances that occur in small amounts in many foods. In all cases they found no substantial differences in composition or amount and therefore concluded that conventional and genetically engineered soybeans are substantially equivalent (Padgette et al. 1996).

Utilizing such studies from the companies that produce genetically engineered products (the FDA does no testing of its own and does not require third-party testing), the FDA formulated its policy on genetically modified foods in 1992, a policy that it still holds to in 2007: "The agency is not aware of any information showing that foods derived by these new methods differ from other foods in any meaningful or uniform way, or that, as a class, foods developed by the new techniques present any different or greater safety concern than foods developed by traditional plant breeding. For this reason, the agency does not believe that the method of development of a new plant variety is normally material information . . . and would not usually be required to be disclosed in labeling for the food" (*Federal Register* 54, no. 104 [1992]: 22991).

This policy narrowly couples labeling of GM foods with safety, leaving out all the other criteria for labels we describe above. The idea is: if there is no safety issue, there is no reason to label. And according to the FDA there is no safety issue, because GM food and traditional food are substantially equivalent, and since traditional food is safe, GM food must also be safe.

Remaining within this tidy argument, the FDA does not hold the method of production to be "material information." This is a telling conclusion that rests on a very specific interpretation of what "material information" means. Normally the term refers to something relevant or pertinent to the matter at hand. But here the FDA has chosen to restrict the meaning, equating "material" with "physical ingredient." Since "genetic engineering," as a method of production, cannot be considered a new ingredient that could call the plant's safety into question, it is deemed irrelevant, and therefore no labeling of GM plants is warranted.

This logic restricts the regulation and labeling of genetically modified foods to the narrowest terms possible. The FDA is applying a very different standard than it does for other foods. Why does the FDA demand labeling of orange juice from concentrate? Surely it is not a safety issue, and just as surely the agency is not raising a question about "substantial equivalence" with fresh-squeezed orange juice, right? When we asked the FDA these questions, the answer was, "No, of course not; it's a matter of truthfulness." A simple and clear answer (from an FDA employee who did not work on GM issues).

Substantial Equivalence Doesn't Tell It All

Two foods are defined as substantially equivalent if the investigation of certain substances in the foods shows the foods to be "the same." The scientists investigate only a select number of known nutrients and antinutrients. Other substances ("nonnutrients") that are produced, expectedly or unexpectedly, through genetic engineering are not taken into account. For example, the fact that each cell of a genetically engineered soybean contains a novel protein—the enzyme that prevents the herbicide from killing the plant—is not included in the data used to determine substantial equivalence. Nor is the fact that genetically engineered plants usually contain five or more DNA sequences that come from other organisms—other plants, bacteria, and viruses (see the description and list on the next page). No traditionally bred plant contains such an array of foreign DNA, which includes, for example, a bacterial gene that gives the plant resistance to a specific antibiotic.

The concept of substantial equivalence is, by virtue of its narrowness, misleading.* The designation suggests that food from a GM crop and from a traditionally bred crop of the same species are the same, and that therefore food from the GM crop is safe. But in reality only a specific subset of substances has been investigated and taken into account in the designation "safe."

Foreign Gene Products as Additives

In the early 1990s, Calgene was developing its "flavrsavr" tomato, the first GM food to enter the market. This tomato was genetically engineered to turn red on the vine but remain hard for shipping. (The idea was that it would taste better, but in the end Calgene's $95 million investment didn't pan out, and the company—on the verge of bankruptcy—was

* For a succinct overview of the problematic concept of substantial equivalence, including links to other articles, see the discussion on the Web site of Physicians and Scientists for Responsible Application of Science and Technology (http://www.psrast.org/subeqow.htm).

What a Genetically Engineered Plant Contains

Plants are not modified by adding just one foreign gene. A whole "DNA construct" made up of DNA from different sources is shot into the plant. In the case of Monsanto's "Roundup Ready" soybeans, which are genetically modified to become resistant to the herbicide glyphosate, the gene construct consists at least of:

- DNA originally from the bacterium *Agrobacterium sp.* (strain CP4) now synthetically produced, for herbicide resistance
- DNA from the cauliflower mosaic virus that regulates the expression of the herbicide resistance gene
- DNA from the petunia to move the gene product to the chloroplasts (so that the herbicide resistance gene will be adequately expressed in the leaves, which are the main target of the herbicide)
- DNA from the bacterium *Agrobacterium tumefaciens* to regulate the production of the enzyme needed for herbicide resistance
- DNA from the intestinal bacterium *Escherichia coli.* The expression of this gene serves as a "marker" to help scientists identify which plants have been genetically transformed in an experiment
- A circular strand of DNA (called a plasmid) derived from the intestinal bacterium *E. coli*

All the other ingredients are biochemically inserted into this last one, the plasmid, which carries the DNA into the plant's cells. When the experiment goes according to plan, every cell of the organism contains at least one copy of the complete construct. Through the gene construct the metabolism of the cells is altered and the plant is obliged to produce novel substances, including an enzyme to convey resistance to herbicide in "Roundup Ready" soybeans and an enzyme affording antibiotic resistance. Other plants might be altered to produce a toxin to kill insect larvae, as is done in "Bt" crops.

In "Roundup Ready" and "Bt" crops, the transgenic proteins are produced continually, in every cell of every plant. By contrast, in normal protein metabolism, proteins are specific to particular tissues and functions. In some cases genetic engineers aim to restrict gene expression to particular tissues by including tissue-specific promoter DNA in the gene construct (for example, in the effort to increase carotene accumulation in canola seeds; see Lindgren et al. 2003).

bought by Monsanto.) The FDA at that time was still in the process of formulating its policy on GM foods. According to Belinda Martineau, a scientist who worked on Calgene's tomato project, "the FDA did not have a specific process in place for dealing with genetically engineered whole foods. It was up to us, therefore, to decide just how to submit to the agency whatever safety data we would produce" (2001, 64).

The company had to decide how to deal with the new tomato's novel proteins. The fruit contained, like other genetically engineered crops, an antibiotic resistance gene and the protein (enzyme) produced through it. The scientists involved came to the conclusion that the enzyme would fit into the category of food additives, and the tomato would then carry a label specifying the additive. Since food additives are strictly regulated, the process of approval might have taken three years. Calgene's management wanted to get around this problem by calling the antibiotic resistance enzyme a "processing aid." Receiving approval for a processing aid was likely to be quicker, the company wouldn't have to publish safety data, and processing aids don't need to appear on labels. This seemed an ideal strategy.

In the end, the FDA did require that Calgene submit a food additive petition, since this was considered the approach that would cause the least problems for approval (Martineau 2001, 161). The main issues considered were whether the tomatoes were safe for human consumption (the company did some animal tests), whether the enzyme might cause resistance to antibiotics in humans, and whether the gene might be transferred to bacteria in the intestines. Based on all the evidence Calgene presented, the FDA declared the enzyme to be a safe food additive (Federal Register 1994).

But it did not require labeling. Normally, the FDA requires any added ingredients to be indicated on a food's label. But in the case of the antibiotic resistance enzyme, the FDA reasoned:

> FDA considers an "ingredient" to be a substance used to fabricate (i.e., manufacture or produce) a food. FDA does not consider those substances that are inherent components of food to be ingredients that must be disclosed in the food label. A genetic substance introduced into a plant by breeding becomes an inherent part of the plant as well as of all foods derived from the plant. Consistent with FDA's general approach on ingredient labeling, the agency has not treated as an ingredient a new

> constituent of a plant introduced by breeding, regardless of the method used to develop the new plant variety. . . . Accordingly, [the antibiotic resistance enzyme is not] an ingredient that must be individually identified on the labels of foods containing them. (Federal Register 1994)

In other words, after declaring the antibiotic resistance enzyme to be an additive (which means it is an ingredient), the FDA immediately turned around and, for purposes of labeling, denied the additive-ingredient status, calling the enzyme an inherent part of the plant. The FDA went on to "solidify" this position by stating that even if the enzyme were an ingredient, it would treat it as a "processing aid" rather than an additive, so it wouldn't require labeling anyway. These logical flip-flops seem inexplicable except as efforts to prevent labeling at all costs. We'll see why later. But now we need to look at how genetic engineering differs from traditional breeding.

A Long Way from Traditional Breeding

It is not only the genetic and biochemical composition that makes GM crops different from conventional crops. The whole way in which they are produced differs radically from traditional breeding, and these differences bring new kinds of uncertainty. First, scientists must isolate the DNA used in the genetic manipulation. This is an involved biochemical and technical procedure, yielding products for which companies (and universities) seek patents. You can't help feeling the wool is being pulled over your eyes when the same companies that argue to the patent office that their "inventions" are "novel" and "non-obvious" in order to receive patents for their "products" also argue to the FDA and others that genetic engineering does not really differ from traditional breeding.*

Traditionally bred plants cannot be patented, although a breeder of a new variety may receive a "Certificate of Protection" from the U.S. Department of Agriculture. This certificate gives the breeder sole rights to sell the variety for eighteen years (Strachan n.d.). But ownership of the variety is limited inasmuch as a farmer using the variety may save the

* That the U.S. Patent and Trademark Office and the Supreme Court have played into the hands of biotechnologists by increasingly broadening what they view as a patentable invention to include genetically modified organisms is another troubling issue, which we will not discuss here (see Krimsky 2003, chapter 4; Andrews 2002; Kaplan 2001).

seeds for his own use and also sell seeds he saves to neighboring farmers. Moreover, the variety may be freely used and exchanged within the scientific research community. In contrast, patents on GM crops give companies the right to forbid farmers from saving and planting GM seeds and to decide whether other scientists are allowed to carry out research with the GM crops.

After isolating the DNA, genetic engineers make a novel gene construct, consisting of DNA from different sources (plants, bacteria, viruses, and sometimes animals [see page 46]). This product of biochemical analysis and synthesis is used to alter the plant, a far cry from a breeder taking pollen from one plant variety and using it to fertilize another plant of the same species in the hope of producing plants with a new combination of characteristics.

Once the laboratory-produced DNA construct has been made in the lab, the task is to get it into the plant. The most prevalent method is to use a "gene gun" (Daniell 1997). Tiny pellets of tungsten or gold are coated with the novel gene construct and then shot into embryonic plants or tissue, which are then grown in a culturing medium. The few that have taken up the target gene can be identified through the expression of a genetic marker that is part of the gene construct. These plants and their progeny are investigated further to see if they also express the desired trait (herbicide resistance, toxin production, and so on). Usually a few plants are chosen to be the parent stock of the future crop. At this point traditional breeding methods set in. Breeders select and perhaps cross the transgenic plants with other known varieties with desirable traits. Finally, a new transgenic crop line emerges.

When a company has a transgenic crop line, it can carry out a detailed genetic analysis to discover how the DNA has been incorporated into the plant. Often they find the DNA-construct is broken, the desired genes are separated from each other, and the individual genes themselves are split into fragments. These genes and fragments may be incorporated into different places in one or more chromosomes. The foreign DNA may also be inserted into a chromosome in such a way that it breaks up one of the plant's genes. Some of the DNA may be broken down altogether. All of this can be detected only after the fact (Makarevitch et al. 2003). And sometimes long after the fact. Monsanto originally declared that each cell of Roundup Ready soybeans did not contain the entire DNA construct (see page 46), but rather a single copy of two of the genes and partial copies of two others. Nonetheless the plant "worked" the way the scientists wanted it to (with high resistance to the glyphosate herbicide). Four years after the Roundup Ready soybeans had been on the market, Monsanto reported that each cell also contained two additional partial segments of the herbicide resistance gene as well as another copy of other DNA from the construct.

Genetic engineering is usually hailed as a precise new technique to make exact modifications. In reality, precision stops when the DNA leaves the laboratory and enters the plant. The scientists have to wait and see what the organism has made of their attempted manipulation.

Unintended Effects

By now you can see why there are good grounds to expect unintended effects arising from any genetic manipulation. The plant may be producing new substances that it hadn't produced before, or its normal production of substances may have been repressed. There are many examples of such changes (Cellini et al. 2004; Freese and Schubert 2004; Wilson et al. 2004).

To make peas more resistant to weevils, scientists in Australia genetically altered peas with a gene construct containing DNA from beans, which have a natural defense against weevils (Prescott et al. 2005). Scientists isolated from beans the gene related to the production of a protein that blocks the breakdown of starch and thereby causes the weevils feeding on the beans to starve. The genetically altered peas had the same kind of defense against weevils. To see if the transgenic peas might present a risk to human health, the peas were tested (in Australia) on mice. To their surprise, the scientists found that the mice developed an immune response to the peas, meaning they produced antibodies against the genetically modified protein. The scientists discovered that the protein in the pea had, after initial formation, been altered (post-translational modification) and in this changed form elicited an inflammatory reaction in the mice. Australia's Commonwealth Scientific and Industrial Research Organization, which carried out the research, decided on the basis of these results not to pursue further work with these GM peas.

Although many undesired and unexpected effects are weeded out in the process of selecting plants for further breeding, commercialized crops can also reveal changes in the field. Some farmers in Georgia complained about the poor performance of their transgenic herbicide-resistant soybeans under conditions of drought and heat. Scientists then carried out a comparative laboratory study of transgenic and conventional soybeans (Gertz et al. 1999). They found that the transgenic plants were shorter, had a lower fresh weight, had less chlorophyll content, and, at high temperature, suffered from stem splitting.

In another case, genetically engineered Bt corn, which produces its own pesticide, was found—after five years of commercial planting on millions of acres—to contain substantially more lignin in the stalks than unmodified corn (Saxena and Stotzky 2001; Poerschmann et al. 2005). Lignin makes the stems woodier. No one has investigated how the higher lignin content might affect the corn's digestibility by the cattle or pigs that are fed on it. What other biochemical pathways may have been repressed or altered as a result of the increased production of lignin also remains unknown.

So genetic modifications intended to change specific and clearly circumscribed characteristics of plants end up affecting the whole organism. Some of the unintended effects may be induced by environmental conditions. No one can foretell the kind or degree of such changes.*

Why a Double Standard?

All these examples make one wonder how the FDA can still hold to its 1992 policy, claiming that it "is not aware of any information showing that foods derived by these new methods differ from other foods in any meaningful or uniform way." This view has a glimmer of credibility only as long as the FDA views GM food through the monocle of substantial equivalence, while wearing blinders to all other considerations. The broader view shows that genetic engineering is a radically new way of altering the plants we utilize for food. Traditional food processing and use of additives begins after the plants have been harvested and reach the factory. With genetic engineering, processing and adding new substances begins already in the growing plant.

Even apart from safety issues, shouldn't the FDA, as a consumer protection agency, inform consumers via labels about genetic engineering as a new method of adding substances to and processing food? Isn't

* Currently we are collecting from the primary scientific literature many examples of the unintended side effects of genetic engineering experiments. To view this collection of examples and commentaries on them visit: http://www.natureinstitute.org/nontarget/index.htm.

this at least as important as knowing that your orange juice is from concentrate? Surely the criteria of truthfulness and honest disclosure remain the same in both cases. Why has the FDA chosen to establish such a double standard?

When the FDA was formulating its policy regarding genetically modified foods, scientists within the FDA itself were critical of the new policy. These dissenting voices were made public only much later. Early in 1992, Linda Kahl, an FDA scientist and compliance officer, complained to James Maryanski, the FDA's biotechnology coordinator, about how the agency was "trying to force an ultimate conclusion that there is no difference between foods modified by genetic engineering and foods modified by traditional breeding practices," which amounted to "trying to fit a square peg into a round hole" (see Alliance for Biointegrity, n.d.). Another FDA scientist, Louis Pribyl, writes in his comments on the draft policy document: "What has happened to scientific elements of this document? Without a sound scientific base to rest on, this becomes a broad, general, 'What do I have to do to avoid trouble'-type document. . . . This document reads like a biotech REDBOOK!! . . . It reads very pro-industry, especially in the area of unintended effects, but contains very little input from consumers and only a few answers for their concerns" (see Alliance for Biointegrity, n.d.).

If the FDA took the perspective of the biotech industry in formulating its policy, it was not alone. Dan Glickman, reflecting on his tenure as Clinton's secretary of agriculture, stated: "Regulators even viewed themselves as cheerleaders for biotechnology. It was viewed as science marching forward, and anyone who wasn't marching forward was a Luddite" (*Los Angeles Times,* July 1, 2001).

The interviewer goes on to write that Glickman expressed his regrets "that industry was allowed to take the lead, as regulators ceded their watchdog role."

This pro-biotech bias makes clear why the FDA policy regarding GM foods differs in spirit and in content so radically from its policies regarding other foods. Otherwise it is impossible to understand the tortuous and often disingenuous argumentation that the FDA uses in its attempts, as Kahl put it, to fit square pegs into round holes. The policy was crafted around a foregone conclusion that is also the biotech industry's view—a GM food should not be labeled. So the FDA had to find ways to narrow the context (safety as the only reason to label) and to "demonstrate" that GM food and food from traditionally bred plants are the same. Along the way it left out all the factually existing differences in content and process, while also ignoring the consumer.

The Consumer's Right to Know?—Contrasting Views

In a panel discussion in 2002 with representatives from the biotech industry and the government, the representative from the FDA opined that "consumers have a right to know—but not to know everything." This echoes a sentiment expressed on the FDA's Web site, where we can read that law "does not require disclosure of information solely on the basis of consumers' desire to know" (FDA 1995). This is a very strange reading of its mission as a consumer protection agency, which is (according to the Food, Drug, and Cosmetic Act) to "promote honesty and fair dealing with consumers."

The Europeans have taken a very different stance on labeling genetically modified food. David Byrne, European Commissioner for Health and Consumer Protection, spoke to the National Press Club on October 9, 2001:

> Let me be very frank. Unless we can give EU consumers confidence in this new technology then GM is dead in Europe. Let me assure you that this is not a scare tactic on my part. I am not prone to exaggeration. . . . As part of the new approval process, GM food and feed will have to be labeled as such. . . . Europe is perfectly entitled to impose the labeling rules proposed. Our consumers are demanding this. They are entitled to choice and full information is now a *right* since the Amsterdam Treaty has become part of the constitutional arrangements of the European Union. . . . Labels that cover all GM-derived products ensure that our consumers are able to choose a GM product or a non-GM product.

This is a very clear position statement. It leaves no doubt that the main reason to label GM foods is the consumer's right to know and to choose. Britain's premier scientific society, the Royal Society, states in its essentially pro–genetic engineering position paper that nonetheless "public debate must take account of wider issues than science alone" (The Royal Society 2002).

In contrast, U.S. regulations (or rather, the lack thereof) concerning

GM food are based solely on a partial examination of the end product. This is called a "science-based" approach. Biotech advocate Henry Miller, of Stanford University's Hoover Institution, echoes the FDA's view when he states that regulations focusing on process are unscientific and that therefore the only "rational approach" is *not* to label genetically modified food (Miller 1999). It is "rational" to ignore process and to ignore the consumer. Following this approach the food label would reveal nothing more and nothing less than what passed through the filter of an extremely narrow view claiming to be "sound science." In the name of this science the government and the biotech industry will choose what is fit for the people to know. This does not sound like a healthy democracy.

Food Labels—A Window to Food's Story

During the course of the twentieth century, farming and food production and processing became increasingly detached from the lives of most people in the Western world. Children who are told, to their surprise, that milk comes from cows and eggs from chickens often find the facts "gross." More broadly, most of us have next to no idea where the food we eat comes from or what is involved in its processing. Food additives labeled on packages are just names. And in many instances we would not be enthralled if we knew more about the food we're eating.

Michael Pollan has written an exemplary piece on the life of a beef steer (2002). His detailed "biography" of the steer brings home how the food we eat is connected with a whole world of social, economic, political, and ecological conditions. When we buy food we're supporting that particular world—a world about which we usually know very little. In a sense we're sleepwalking. This is one of the consequences of the technologization of our culture—it distances us from concrete processes, and we end up living in a world peopled with end products whose life stories we don't know.

Food labels are one (and only one) small window into the world of food. Many labels restrict themselves to end product information—contents, nutritional value, etc. But the more they tell you about processing and method of production, the more you as a consumer can see into the world connected to that particular product. Some producers want you to have this knowledge, especially in the natural foods industry, where the label might tell you that the contents of salsa are organically grown (and therefore that genetic engineering was not involved in the food's production). Or the milk you buy might state that the cattle were not treated with hormones and did have access to pasture grazing. All this labeling is voluntary. What you don't find (although it is allowed) is voluntary labeling that states "genetically engineered" or "factory farmed." Evidently the producers are not interested in any window being open to those worlds.

The more the window to a food's story is open and the larger that window is, the greater the opportunity for consumers to make informed decisions. In respect to genetically engineered foods (and to many practices of modern agriculture and animal husbandry) the government keeps the window closed. It does not go all the way to promote "honest and fair dealing with consumers."

I doubt that the impetus for making labels transparent will come from the government. The pro-biotech stance is very firm. But if consumers become more active—as they have in Europe—then things can change. This activity can have at least two complementary directions.

First, we can inform the government that we do believe in the significance of our choices as consumers and in the obligation of the federal regulatory agencies to protect these choices. We can demand more open and consistent practices instead of double standards. We can demand that labels be more comprehensive and include information about process and production as well as content.

The cynic among us will counter that we don't have any power to confront the government-biotech complex. But the consumer does still have power. When the Agriculture Department was adopting new "organic" standards, public protest brought about changes so that, for example, a product cannot be called organic if it has been genetically engineered. The standards on organic agriculture are far from perfect, but they are much better than they would have been had citizens not made their opinions known.*

Second, we can purchase selectively. We can choose to buy food that does open the window to food's story. In this way we influence the way food is produced and labeled. The more consumers buy products with transparent labeling practices, the more such labeling—and the type of farming and animal husbandry the labels stand for—will take hold.

---

* For information about how you can make your opinion known, see the Web sites for The Campaign (http://www.thecampaign.org) and the Center for Food Safety (http://www.centerforfoodsafety.org).

Consumers are a primary force in the rapid expansion of the organic foods market, as well as in the burgeoning growth of local and regional food networks, such as Community Supported Agriculture (CSA), where there is direct contact between consumers and farmers who produce the food.

Surveys consistently show a majority of Americans in favor of labeling GM foods.* While the FDA has, for now, set itself against full disclosure, surely it cannot remain wholly immune to consumer pressure. After all, if the FDA does not exist to protect the declared interests of consumers, particularly with respect to truthfulness and transparency, then what *is* its mission?

---

* See, for example, http://www.harrisinteractive.com/harris_poll/index.asp?PID=478 and Hallman et al. 2004.

Part II

Genes and Context

Chapter 5

Genes Are Not Immune to Context

Examples from Bacteria

One of the most widespread misconceptions concerning the nature of genes is that they have a defined and fixed function that allows them to operate the same in all organisms and environments. We have a picture of robust genes determining all the characteristics an organism has. And any given gene will do the same thing in a bacterium as in a corn plant or human being. It doesn't care where it is. The gene bears its set of instructions wherever it goes and strictly carries out its duty.

This picture informs genetic engineering. Take a gene from bacteria and put it into a plant, and the plant will produce its own pesticide or become resistant to an herbicide. Since such transgenic plants exist, the proof is evidently in the pudding. Genetic manipulation works; genes are faithful workhorses. But does genetic manipulation work the way we imagine with our schematic pictures? What else may be occurring that doesn't fit into such a neat mechanistic scheme?

It's somewhat ironic that precisely within the last ten to fifteen years—the period in which genetically modified crops have been developed and commercialized in the United States and some other countries—a wealth of research on genes in relation to environmental effects has been carried out, showing that genes are anything but automatic instruction sets immune to their context. This research has significant implications for the way we assess genetic engineering. Unfortunately, it often seems that the results of this basic research have little effect on the minds and pocketbooks supporting the global drive to manipulate organisms genetically. In this chapter we discuss some examples of the contextual gene in bacteria.

The Interactive Gene

With the widespread use of antibiotics in our culture, many bacteria have become resistant. They thrive even when subjected to high doses of antibiotics. As a rule, the resistance comes at a cost, since the resistant bacteria tend to grow slowly. But when they are grown in laboratory cultures, some of these resistant bacteria will experience so-called compensatory mutations—they stay resistant, but change genetically in a way that allows them to grow fast like wild, nonresistant strains. Others mutate back to the wild form and lose their resistance altogether.

The question arises whether such mutations (changes in genes or in higher-order genetic structures) are in any way dependent on the environment. The traditional view, rooted deeply in the neo-Darwinian theory of evolution, holds that genes mutate spontaneously and independently of the environment. The classical experiment with bacteria by Salvadore Luria and Max Delbrück in the 1940s gave clear evidence that such spontaneous, milieu-independent mutations exist (Luria and Delbrück 1943). For decades this experiment (along with other evidence) served as the rock-solid "proof" that genetic mutations, except for extreme cases involving irradiation or exposure to chemical toxins, are not influenced by their environment. But more recent research shows that mutations do in fact arise in response to changing environmental conditions.

A group of biologists in Sweden investigated whether the above-mentioned compensatory mutations and the reversion to the wild form in bacteria are influenced by the environment (Björkman et al. 2000). They grew antibiotic-resistant bacteria—in the absence of antibiotics—as laboratory cultures (in petri dishes) and also inoculated mice with the same bacteria. The researchers monitored the mutations that occurred in the bacteria in these two different habitats. They found that compensatory mutations occurred in both habitats, but, to their surprise, they discovered that the way the genetic material changed differed significantly depending upon the environment. In the case of streptomycin-resistant bacteria in mice, they found ten cases of identical compensatory mutations *within* the resistance gene. In contrast, this gene never mutated in the lab-cultured bacteria, where they found fourteen compensatory mutations in genes *outside* the resistance gene. Evidently, the environment had everything to do with what kind of mutations occurred. "Mice are not furry petri dishes," as the title of a commentary article put it (Bull and Levin 2000).

The authors conclude that the mutations are "condition-dependent" and suggest that some unknown "mutational mechanism" limited the mutations in the mice to a specific part of the resistance gene while also increasing its mutation rate. Whatever the details of cell physiology turn out to be, it is clear that the genome of the bacteria changes in relation to a specific kind of environment. The bacteria—down into their genes—interact with and evolve in relation to their environment.

Adaptive Mutations

In another study (Bjedov et al. 2003), a research group in France gathered wild strains of the bacterium *E. coli* from a wide variety of environments—the large intestines of humans and different animals, soil, air, and water. In the end they collected 787 different strains. These strains were given optimal conditions in lab cultures and began to grow and multiply rapidly, mimicking ideal conditions in nature where bacteria reproduce quickly. But in nature, bacteria are also exposed to times of dearth, where the substrate they live upon, for example, is suddenly used up. To mimic these conditions, the researchers withheld nutrients for a seven-day period. Most bacteria survive under these conditions, but they no longer grow and divide.

The scientists measured the rate of mutations occurring in the cultures the first day after withholding nutrients and then again at the end of the seven-day period. During this time the mutation rate increased on average sevenfold. In other words, the mutation rate increased dramatically when the bacteria no longer received adequate nutrition. The bacteria switch, in the words of the authors, "between high and low mutation rates depending on environmental conditions" (1409).

Such a stress-induced increase in mutation rate has been discovered in laboratory strains of bacteria. (For reviews of this still controversial topic, see Wirz 1998; Rosenberg 2001; Foster 2004; Rosenberg and Hastings 2004; Roth et al. 2006). Does this increase in mutation rate serve the bacteria, or is it a kind of last gasp, a dissolution of the bacteria before they die of starvation? It seems that bacteria produce unique kinds of mutations during such periods of physiological stress, some of which help the bacteria survive specifically under those conditions. One speaks of "adaptive mutations."

For example, there are strains of *E. coli* that have lost the capacity to utilize the sugar lactose as a source of energy. If such a strain is cultured in a starvation medium with lactose as the only energy source, most of the bacteria remain in a stationary phase until they die. But under these conditions some of the bacteria begin to hypermutate, which means they rapidly create a large number of mutations, and among these are ones that allow them to live from lactose. The bacteria with this ability survive, multiply, and form new colonies. In at least some cases such adaptive mutations arise only in the specific medium—that is, the mutations allowing bacteria to utilize lactose don't occur when bacteria are grown in a medium with sugars other than lactose.

In other instances, *E. coli* bacteria do not hypermutate, but find another way to deal with the environmental challenge (Hastings et al. 2000). Some of the bacteria in the medium with lactose produce multiple copies of the gene related to their inability to live from lactose. This gene amplification seems at first absurd. But scientists found that *E. coli* strains unable to grow when they receive only lactose as a nutrient do form enzymes that break down lactose, but in inadequate amounts. When the bacteria amplify the defective lactose enzyme gene, the cumulative effect is that they produce enough enzymes to break down a sufficient amount of lactose to grow slowly and survive—a remarkably active and meaningful genetic adaptation. This amplification occurs in no other genes in the bacteria. It is specific to the lactose enzyme gene and clearly induced by the environment.

Transfer of Resistance

Bacteria have a further way of adapting to new conditions. Antibiotic-resistant bacteria have already been mentioned. Cholera bacteria, for example, are normally susceptible to different antibiotics. After 1993, however, antibiotic-resistant cholera bacteria rapidly spread around the globe. How could this occur? Scientists discovered that these bacteria are simultaneously resistant to at least five different antibiotics. They found that the genes related to this resistance were all grouped together and formed a "packet" of genes that can move from bacterium to bacterium.

A research group at Tufts University in Boston discovered conditions that facilitate this movement and uptake of genes (Beaber et al. 2004). When bacteria are grown in cultures with concentrations of antibiotics that are not sufficient to kill them, they go through physiological changes similar to what happens to bacteria in a starvation medium. Part of this transformation is called an SOS response. It comes about when DNA is damaged and involves DNA repair and duplication. The Tufts scientists found that during the SOS response the bacteria also increased the transfer of the resistance gene clusters to other bacteria. Evidently, the use of antibiotics promotes the spread of antibiotic resistance among bacteria. In this way, once resistance is anchored in mobile genetic elements, it can spread rapidly.

The above examples show how strongly the environment influences the activity of genes, induces changes within genetic structures (mutations), and stimulates the movement of genes between bacteria. Bacteria are in continual interplay with their environment, actively responding to changing conditions. And this responsiveness and flexibility includes genes. If we release genetically engineered bacteria into the environment, there is little doubt that in time they will be passing their genes to other bacteria, as well as receiving genes from other bacteria and mutating according to changing circumstances. Whether the manipulated foreign genes they carry will be exchanged and how they may affect or be affected by the dynamics of genetic responses to changing environments are completely open. But two things we can know for sure: these genes will not function immune to the changing circumstances, and things will happen that no one expects or can foresee. We're not saying this to promote fear, but to dissolve the illusion that we can keep under control what we have released into the world in this way. Genes have their own robust nature, but it is part of this nature to be in interaction with the world.

Chapter 6

The Gene

A Needed Revolution

This short chapter is about the gene and includes many statements about this central concept of modern biology from geneticists and from historians and philosophers of science. The quotes cited here are like footprints, indicating the pathway and evolution of modern genetics. A fascinating biography of a concept emerges. And the results of research in the past few decades have brought biology to a threshold that calls for a long-needed revolution in the way we interpret life.

The concept of the gene was first conceived by Gregor Mendel in the 1860s. He never used the term "gene," but spoke of "factor," "Anlage," or "element" to point to the underlying cause of differences in inherited characteristics of different offspring. He writes, for example: "The distinguishing characteristics of two plants can only be due to the differences in the make-up and grouping of those elements that stand in vital interaction within the germ cells" (Mendel 1866).

In 1909, Danish biologist Wilhelm Johannsen coined the term "gene" to refer to discrete determiners of inherited characteristics: "The word gene is completely free of any hypothesis; it expresses only the evident fact that, in any case, many characteristics of the organism are specified in the germ cells by means of special conditions, foundations, and determiners which are present in unique, separate, and thereby independent ways—in short, precisely what we wish to call genes" (quoted in Portugal and Cohen 1978, 118).

Most people today are familiar with the term "gene" and have learned in school and through the media that genes determine the characteristics of organisms. There are genes for hair and eye color, genes that direct the formation of our body's substances, and many genes that are somehow defective and cause disabilities and illnesses—genes for diabetes, cancer, schizophrenia, and more. Almost no one talks about human, animal, or plant physiology today without ascribing a central role to genes.

This deterministic gene is essentially the gene of the first half of the twentieth century. It is the gene most people have in mind today, more than half a century later. This gene has been described in the following way by different geneticists:

> In a specified environment, genes determine what kind of an individual a representative of a given species is going to be. There can be little doubt that genes also determine to what species a given individual will belong. By logical extension, it can be argued that genes determine whether an organism is a plant or an animal, as well as what kind of a plant or animal. And, to carry these deductions still further, genes determine whether or not an organism is going to develop at all. (Sturtevant and Beadle 1939, 334)

> Mendelian inheritance is essentially atomistic, the heritable qualities of the organism behaving as if they were determined by irreducible particles (we now call them genes). (Horowitz 1956)

> It has been known since about 1913 that the individual active units of heredity—the genes—are strung together in one-dimensional array along the chromosomes, the threadlike bodies in the nucleus of the cell. . . . In recent years it has become apparent that the information-containing part of the chromosomal chain is in most cases a giant molecule of DNA. (Benzer 1962)

The Watson-Crick double helix model of DNA (1953) and subsequent discoveries from the late 1950s into the 1970s relating DNA to protein synthesis provided a mechanistic model of the gene and of gene action that inaugurated the age of molecular biology. This was the time of boundless optimism concerning the ability of the reductionist approach to decipher the mechanism of life. As James Watson states in his classic and influential textbook, *The Molecular Biology of the Gene*: "We have complete confidence that further research of the intensity given to genetics will eventually provide man with the ability to describe with completeness the essential features that constitute life" (quoted in Darnell et al. 1986, 1).

With advances in geneticists' knowledge, gene action has come to be viewed as an increasingly complex process, so that to state what a gene is requires longer statements filled with technical terms that no one but a specialist can understand. Witness the definition in a comprehensive textbook about the gene by Singer and Berg (1991, 622):

> A [eukaryotic] gene is a combination of DNA segments that together constitute an expressible unit, expression leading to the formation of one or more specific functional gene products that may be either RNA molecules or polypeptides. The segments of a gene include (1) the transcribed region (the transcription unit), which encompasses the coding sequences, intervening sequences, any 5' leader and 3' trailer sequences that surround the ends of the coding sequences, and any regulatory segments included in the transcription unit, and (2) the regulatory sequences that flank the transcription unit and are required for specific expression.

But the advances in genetics have not only refined the mechanistic model. The complexity at the molecular level reveals that the simple mechanisms imagined in the 1960s simply do not exist in that form. It has become less and less clear what a gene actually is and does. And although the deterministic gene is still the gene that lives in the minds of many students, lay people, and—at least as a desire—in the minds of many biologists, the findings of late-twentieth-century genetics show one thing clearly: the simple deterministic gene, the foundational "atom" of biology, is dead. There is no clear-cut hereditary mechanism—no definite sequence of nitrogenous bases in a segment of a DNA molecule that determines the make-up and structure of proteins, which in turn determine a definite feature of an organism.

What follows is a series of statements about the contemporary gene—the gene of the past two decades. This gene looks very different from the one described above:

> The more molecular biologists learn about genes, the less sure they seem to become of what a gene really is. Knowledge about

the structure and functioning of genes abounds, but also, the gene has become curiously intangible. Now it seems that a cell's enzymes are capable of actively manipulating DNA to do this or that. A genome consists largely of semistable genetic elements that may be rearranged or even moved around in the genome thus modifying the information content of DNA. Bits of DNA may be induced to share in the coding for different functional units in response to the organism's environment. All this makes a gene's demarcation largely dependent on the cell's regulatory apparatus. Rather than ultimate factors, genes begin to look like hardly definable temporary products of a cell's physiology. Often they have become amorphous entities of unclear existence ready to vanish into the genomic or developmental background at any time. (Historian of science Peter Beurton, geneticist Raphael Falk, and historian of science Hans-Jörg Rheinsberger, 2000)

The gene is no longer a fixed point on the chromosome . . . producing a single messenger RNA. Rather, most eukaryotic genes consist of split DNA sequences, often producing more than one mRNA by means of complex promoters and/or alternative splicing. Furthermore, DNA sequences are movable in certain respects, and proteins produced by a single gene are processed into their constituent parts. Moreover, in certain cases the primary transcript is edited before translation, using information from different genetic units and thereby demolishing the one-to-one correspondence between gene and messenger RNA. Finally, the occurrence of nested genes invalidates the simpler and earlier idea of the linear arrangement of genes in the linkage group, and gene assembly similarly confutes the idea of a simple one-to-one correspondence between the gene as the unit of transmission and of genetic function. (Geneticist Peter Portin, 1993)

Whether a particular gene is perceived to be a major gene, a minor gene or even a neutral gene depends entirely on the genetic background in which it occurs, and this apparent attribute of a gene can change rapidly in the course of selection on the phenotype. (Developmental biologists H. Frederik Nijhout and Susan M. Paulsen, 1997)

The preceding descriptions point to the contextual nature of the gene: if you "have" a gene at one point in time, it may become, both structurally and functionally, something quite different at another time or place. As a result, it is no longer possible to speak of the gene in a straightforward manner:

> There is a fact of the matter about the structure of DNA, but there is no single fact of the matter about what the gene is. [Genetics today] provides strong, concrete support for the claim that the concept of the gene is open rather than closed with respect to both its reference potential and its reference. (Philosopher of science Richard M. Burian, 1985)

> Paradoxically, in spite of the new, sometimes overwhelming, concreteness of our comprehension of the gene as a result of DNA technology, we seem to be left with a rather abstract and generalized concept of the gene that has quite different significances in different contexts. . . . It should, however, be strongly emphasized that our comprehension of the very concept of the gene has always been abstract and open as indicated already by Johannsen [in 1909]. (Geneticist Peter Portin, 1993)

> [In the molecular gene concept] "gene" denotes the recurring process that leads to the temporally and spatially regulated expression of a particular polypeptide product . . . the gene is identified not with these DNA sequences alone but rather with a process in whose context these sequences take on a definite meaning. (Philosopher of science Paul E. Griffiths and biologist/philosopher of science Eva M. Neumann-Held, 1999)

Because the gene has become something so very different from the clearly circumscribed determinant it started out as, some geneticists think it is time to leave it behind:

> For biological research, the twentieth century has arguably been the century of the gene. The central importance of the gene as a unity of inheritance and function has been crucial to our present understanding of many biological phenomena. Nonetheless, we may well have come to the point where the use of the

> term "gene" is of limited value and might in fact be a hindrance to our understanding of the genome. Although this may sound heretical, especially coming from a card-carrying geneticist, it reflects the fact that, unlike chromosomes, genes are not physical objects but are merely concepts that have acquired a great deal of historic baggage over the past decades. (Geneticist William Gelbart, 1998)

> Our knowledge of the structure and function of the genetic material has outgrown the terminology traditionally used to describe it. It is arguable that the old term gene, essential at an earlier stage of the analysis, is no longer useful, except as a handy and versatile expression, the meaning of which is determined by the context. (Geneticist Peter Portin, 1993)

The gene concept is unlikely to be discarded, since it is far too deeply entrenched in the minds of scientists and the public. But we need to realize that the popular usage of the term, which still implicates the gene as the definitive causative agent in biology, simply does not coincide with biological reality.

As geneticist Peter Portin remarks in one of the above quotes, "the very concept of the gene has always been abstract." In other words, the gene is not a thing at all, but a way of ordering and interpreting phenomena. This may be surprising to anyone used to thinking about genes as concrete biological substances that make things happen. The gene as a robust "thing" is a figment in the materialist mind, a mind that can only conceive the world as governed by mindless material entities that (somehow) carry out meaningful processes.

I do not want to suggest that the concept of the gene has no relation to material happenings. But the gene concept was not, in the first place, derived from engagement in the richness of hereditary phenomena. It was a preconceived notion that framed scientists' thinking and action. Experiments were designed with the gene concept in mind, and investigators then interpreted the results in terms of the particulate conception of inheritance they presupposed in the first place. In the best case (for example, Mendel's experiments with peas or many experiments in the early twentieth century with the fruit fly), experiments showed a partial fit with the conceptual framework. Researchers homed in on the fit and delved ever more into biological minutiae. The gene concept opened up worlds and seemed to be supported by a great number of experiments.

As different researchers pursued a variety of directions of inquiry, the phenomena at the molecular level showed increasing complexity and variation. As a result, any schematic representation of the gene just didn't work, and a colorful array of definitions of the gene emerged, as the above quotes show.* In view of the plethora of gene definitions, philosopher of science Philip Kitcher concludes: "A gene is anything a competent biologist has chosen to call a gene" (1992).

This statement does not indicate a fall into total relativism. It is simply the indirect acknowledgment on the part of contemporary genetics that there is no particular *this* (gene) determining a particular *that* (trait). So to retain a connection to the actual phenomena, geneticists have come to describe the gene as a potential, as a process, and as dependent on the organismic context. In other words, the mechanistic conception of the gene as a power unto itself, elevated above the turmoil and complexity of day-to-day cellular life and doing its thing under any and all conditions, has to be discarded. We cannot satisfactorily adapt the static gene concept to the dynamic reality of the organism.

A great gift of recent genetic studies is that they show in a rich and varied way at the microlevel what we could have known all along from a study of organismic life at the macrolevel if our minds had been open: every organism develops from an open potential and forms over time in dynamic interaction with its own developmental process and its (changing) environment. Only insofar as the mechanistic paradigm holds the human mind captive do we come to think of and believe in genes as neatly circumscribed material determinants.

The gene is an abstraction—a product of a process of isolation, as the eminent neurologist Kurt Goldstein would have said—that has guided the development of genetics for over a century. The idea of a fundamental unit of inheritance, the idea of the grand mechanism that determines life, a mechanism that the human mind can fathom and eventually control, has fired the minds of modern geneticists.

But the research itself—the immersion in the phenomena mined from living organisms via experimentation—brings scientists and their

* The history of genetics, from the early twentieth century on, provides many examples of observations and experimental results that did not fit the dominant gene paradigm. But only within the past couple of decades has the evidence become so glaring that it can no longer be ignored by the scientific community.

concept of the gene to a boundary. It is a boundary one can ignore, as is largely the case in commercialized genetic engineering. It is a boundary that can stimulate scientists to tweak existing models to better fit experimental results. But it is a boundary that can also be felt existentially and become a stimulus for a mental and methodological revolution:

- Can we take reality so seriously that we actually give up—in our heart of hearts and in our innermost thought forms—rigid conceptions like that of the gene?
- Can we do without the security of a guiding notion that imagines discrete entities working in chains of cause and effect to constitute the stuff of life?
- Can we get beyond the "thing" mind-set altogether, which is informed by fixed concepts, and learn to consciously swim in and adapt ourselves to a new medium, namely the fluidity and dynamics of the organic world?

These are radical questions. If we answer them with "yes" and our swimming exercises begin in earnest, we will encounter wholly new facets of the world. It seems that the phenomena themselves are calling for this revolution.

Chapter 7

Reflections on the Human Genome Project

During the 1990s molecular biologists were fully engaged in a race to determine the complete DNA sequence in various organisms. And they succeeded—first in bacteria, then in yeast, and finally in a well-researched roundworm (*C. elegans*). In early 2000 the DNA sequence of the fruit fly, the genetic workhorse of the twentieth century, was completed. In June 2000, at the White House amid media fanfare, two genome sequencing teams announced that they had completed a "working draft" of the human genome. Their reports were published in February 2001 (International Human Genome Sequencing Consortium 2001; Venter et al. 2001). The mega-project was at an end—or was it actually just the beginning?

Another Century of Work

In 1991 geneticist Walter Gilbert made a brash statement: "I expect that sequence data for all model organisms and half of the total knowledge of the human organism will be available in five to seven years, and all of it by the end of the decade" (Gilbert 1991). With regard to sequencing, Gilbert was astoundingly close in his conjecture. At that time almost no one believed the feat could be accomplished in only ten years. But technical advances in automated, rapid sequencing, along with more powerful supercomputers and software, helped accelerate the genome work. The competition between the two genome teams, one privately and one publicly funded, was also a major driving factor.

But Gilbert saw more in the sequence completion than virtually

---

This chapter was cowritten with Johannes Wirz.

endless strings of letters on a computer screen, representing nitrogenous bases in DNA. He spoke of gaining "total knowledge of the human organism." This statement reflects a tendency—one that seemed to accelerate in stride with gene-finding—to make overblown claims about the genome work. We might expect such hyperbole from the media seeking the hottest stories, but the scientists involved in the work were often the worst transgressors of measured assessment. The genome project was, in the words of the publicly funded team's leader, Francis Collins, "the most important and most significant project that humankind has ever mounted" (quoted in Kolata 1993). Why? Because it meant opening what he, like many others, called "the Book of Life," a book that reveals the secrets of the human being. "For the first time," stated biologist Robert Weintute, "we are reducing ourselves down to DNA sequences . . . to rather banal biochemical explanations. . . . We are dealing with the mystery of the human spirit" (quoted in Wade 1998).

When in 2000 a *New York Times* headline announced, "Genetic Code of Human Life Is Cracked by Scientists," the lead article proclaimed: "In an achievement that represents a pinnacle of human self-knowledge, two rival groups of scientists said today that they had deciphered the hereditary script, the set of instructions that defines the human organism" (Wade 2000). Interestingly, at this pinnacle of fervor concerning the project, some scientists were markedly more circumspect in their comments. Molecular biologist David Baltimore remarked, "We've got another century of work to figure out how all these things relate to each other" (quoted in Angier 2000). A year later, Svante Pääbo, another leading genome scientist, remarked on the "insidious tendency to look to our genes for most aspects of our 'humanness,' and to forget that the genome is but an internal scaffold for our existence" (Pääbo 2001). And still another geneticist stated, "It's like a book in a foreign language that you don't understand. That's the first job, working out the language" (quoted in Pennisi 2001).

These scientists are telling us that the genome project was actually just the beginning of real understanding. It is, after all, one thing to find a scaffold or a book that you haven't even begun to decipher (and we should remember, in applying the book metaphor, that possessing a physical book with its text does not make us masters of the text). It is a wholly different matter to gain knowledge of the actual workings of the living organism, not to mention self-knowledge and a key to "the mystery of the human spirit."

So, was the genome project just caught up in one big jamboree of hype? In many ways, yes. In a letter to the editor of *Nature,* written before the completion of sequencing was announced, scientist Sol Hadden put his finger on some essential issues:

> Current hype about the expected completion of the Human Genome Project demands some clarification. Although initially conceptualized more broadly, the project is effectively about determining the sequence of bases in the human genome. This is not the same as trying to understand the program that is encoded in human DNA. Consequently, the results will be in the merely descriptive naturalistic tradition. Technical development has always had that effect on scientific disciplines, for example the electron microscope, the radio telescope or the automated DNA sequencer.
>
> Of course, researchers are always quick to emphasize the importance of their work to whatever application is in vogue, and curing disease is a worthy goal. But how will the Human Genome Project help to achieve this end? A look at any [gene map] from any species reveals what looks like an explosion in a slaughterhouse. Where is the order we need, to make sensible rather than trial-and-error genetic manipulations?
>
> In any case, pharmacogenomics [using genetics to make medicines] requires an understanding of the apparent genetic "disorder" in any organism's genome, of genotype-phenotype mapping, of gene-gene interactions, of intraspecific genetic variability, and of self-organizational processes, rather than endless lists of DNA bases. (Hadden 2000)

In other words, the Human Genome Project really serves to show how little we know. And we could have realized all along—if hype did not have such a strong pull on us—that reams of data (2000 New York City telephone books' worth) would not tell us much. The real challenge is to understand genes in the context of the living organism and not to connect this endeavor with the expectation that such knowledge will open up the secrets of life.

Only 20,000 Genes?

During the 1990s, scientists believed that there must be approximately 100,000 protein-coding human genes. One of the most surprising conclusions that both genome sequencing teams drew from their data was that the human genome contains only about 30,000 genes (International Human Genome Sequencing Consortium 2001; Venter 2001). In 2004 the International Human Genome Sequencing Consortium reduced the number of genes to a range between 20,000 and 25,000. Reporting at a 2007 meeting on the Biology of Genomes, computational biologist Michele Clamp said that the number of genes is still "a mess," since the three main databases on protein-coding genes don't agree with one another. Her own analysis led her to the for-the-present definitive number of 20,488 protein-coding human genes (cited in Pennisi 2007).

While scientists have begun to get acclimatized to the idea of such a small number of human genes, the low number was unexpected because far less complex organisms have nearly as many protein-coding genes. The roundworm (consisting of a total of 959 cells!) has about 20,000 genes, while the mustard plant *Arabidopsis* has about 25,000. If, as the story goes, genes make an organism, how can it be that we—with our complex internal organs and physiology, not to mention behavior—have such a small number of genes?

The real question is, however, why did anyone think that genes make an organism what it is in the first place? As Svante Pääbo comments, successes in the last decade "have resulted in a sharp shift toward an almost completely genetic view of ourselves. I find it striking that 10 years ago, a geneticist had to defend the idea that not only the environment but also genes shape human development. Today, one feels compelled to stress that there is a large environmental component to common diseases, behavior, and personality traits! There is an insidious tendency to look to our genes for most aspects of our 'humanness,' and to forget that the genome is but an internal scaffold for our existence" (Pääbo 2001).

What is so strange about the genocentric view is the fact that the genetic discoveries themselves don't actually support it. The results are simply being viewed through a deterministic and materialistic lens.

Genes and Development

During the past fifteen years the role of genes in development has been studied intensively and can help shed light on the relation between an organism and its genes. In 1994, Walter Gehring's research group in

Basle, Switzerland, discovered that the human being, mouse, and fruit fly all have a gene—called Pax 6—related to eye formation that is very similar (homologous) in all three species (Quiring et al. 1994). This came as a surprise, since the eyes of mammals and insects are totally different anatomically. No one expected the "same" gene to be related to such different structures.

The apparent connection between the Pax 6 gene and eye development became more compelling when researchers were able to manipulate fruit flies to express the Pax 6 gene in tissues that would normally become wings, legs, and antennae (Halder et al. 1995). The result was wholly abnormal fruit flies with partial eyes growing on their legs and wings and even on their antennae. However, these parts often did not develop fully. The scientists then proceeded to do the same experiment with the homologous Pax 6 gene from the mouse. The fruit flies again made eyes—fruit fly–type and not mouse-type—on other body parts. The same experiment succeeded with Pax 6 genes from sea squirts and squids. Gehring concluded that they had clearly discovered and demonstrated the existence of a "master control gene" for eye development (Halder et al. 1995; Gehring and Ikeo 1999).

But, as is usually the case in biology, the story and the conclusion are not so straightforward. Since the Pax 6 gene is in yet unknown ways functional in animals without eyes, like roundworms and sea squirts, it is clearly not related to eye development in these organisms. In other organisms it is also connected to different developmental processes. Mutant mice with two copies of the altered Pax 6 gene not only have no eyes at all, but they have malformed noses, cannot breathe, and die. In squids the gene is active in tentacle formation. In the fruit fly it is involved in the development of other parts of the nervous system besides the eye, and if the Pax 6 gene is not expressed at all in mutants, they die. And in the fishlike lancelets (*Branchiostoma*), it is related to the development of olfactory and central nervous system tissue.

So it seems that, in each organism where it has been found, the "master control gene" for eye development is involved in processes other than eye development. Within a particular organism it is active at different places and at different times, depending on the organ or tissue that is forming there and then (see figure 7.1). Evidently, it's not just the gene that determines the function.

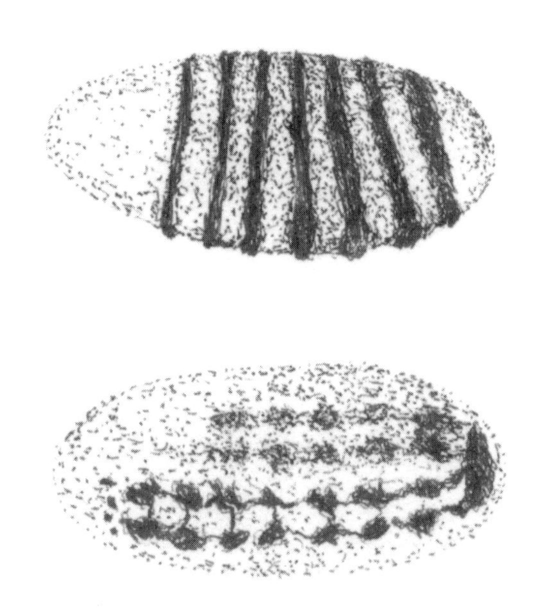

Figure 7.1. One gene, different functions. The FTZ gene in the fruit fly is needed to form a particular protein (the fushi tarazu protein). But the gene and this protein have more than one function during the fly's embryonic development. The drawings show two fruit fly embryos, one at an earlier stage of development (top), the other at a later stage (bottom). The dark stripes and blotches represent the FTZ protein, which was made visible by staining. In the earlier stage (top), this protein is expressed in bands and is active in the formation of segment boundaries; it is then broken down. Only three hours later (bottom), the protein is formed anew and is involved in the development of nerve cells. Thus the FTZ gene is first a "gene for" segment development and then a "gene for" nerve cell development. (Drawing by Craig Holdrege, based on photograph in Duboule and Wilkins, 1998)

The Resourceful Organism

One finds many examples in the study of developmental genes that unfold like this: first a gene is discovered in a particular organism within a particular experimental and developmental context. Then this "same" gene is discovered in other organisms and usually has at least some similar functions. The more the gene is researched, the more it turns out to be implicated in various developmental processes. In the end, the "same" gene has neither a common function among different species, nor only one function within a single species.

This fact led Denis Duboule and Adam Wilkins to use the term "bricolage" to express how the organism uses what is genetically at hand to realize its own specific development. They expect that "the primary source of developmental differences between fruit flies and foxes will

Figure 7.2. The lancelet (*Branchiostoma*) is a fishlike animal that dwells in coastal waters and burrows into sand. About two inches long, it feeds by straining small organisms out of the water. (Drawing by Craig Holdrege)

prove to be not unique genes but rather the way that comparable, or the same, gene functions are differently deployed in their development" (Duboule and Wilkins 1998).

An experiment illustrates this fact clearly (Manzanares et al. 2000). The lancelet (*Branchiostoma*) is a close relative to the vertebrates and is often used to depict how the evolutionary ancestor of vertebrates might have appeared. It is a small fishlike creature that has, however, no bony skeleton and no paired fins (see figure 7.2). Its front end is pointed, and biologists don't speak of a head because typical head features, like a brain and brain capsule, developed sense organs, or a jaw, are missing.

Scientists have found a group of developmental genes, called the Hox genes, that are related, among other things, to the formation of head structures in vertebrates. These Hox genes were also discovered in the lancelet, and since it is has no head, these genes must be related to other, up till now unknown, processes in lancelet development. When, however, the sequences that regulate lancelet Hox gene expression were implanted into mice and chick embryos, they turned out to influence genes in head-forming tissues. This means that a DNA sequence with specific functions in one organism can be utilized by another organism to form completely different tissues and organs.

Both this and the "eye" gene example show us that genes don't make the organism. What a gene "is" is dependent on the organism in its spatially and temporally unfolding existence. You always have to presuppose the organism to understand the gene. This conclusion has far-reaching implications.

Take, for example, our conception of evolutionary processes. According to the scenario taught in schools and universities worldwide, the gradual accumulation of gene mutations causes organisms to evolve new characteristics. But this scenario doesn't work if we take the results of developmental genetics seriously. Rather, we must imagine the evolving organism utilizing "old" genes in new ways to realize new evolutionary developmental characteristics. This view removes genes from their pedestal in evolutionary theory, since they can no longer be seen as the driving evolutionary force. The whole organism—which has been virtually lost in genetic and evolutionary thinking today—returns to the center stage of development and evolution (see also Holdrege 2005b).

Genes and Human Traits

> The work on viruses and bacterial cells that gave birth to molecular biology in the 1940s and 1950s significantly strengthened the earlier Mendelian notion of "one gene, one function." Furthermore, in recent decades, geneticists and molecular biologists have inadvertently contributed to this misconception by the ways they name their genes based on how they were first identified—breast cancer genes, growth factor genes, and so on. This semantic imprecision has had an unfortunate effect on public perception of gene action: many lay people apparently believe that phenotypic traits, such as blue eyes or obesity, are due to the exclusive function of particular genes. . . . Explicit recognition of the general rule of multiple use of specific regulatory gene products would help to clarify issues in both development and evolution. (Duboule and Wilkins 1998, 56)

Just as the 1990s were the decade of genome sequencing, so also were they the decade in which hardly a day went by without an announcement of the discovery of a new gene determining some trait: longevity, happiness, day-night rhythm, alcoholism, schizophrenia, sex drive, Alzheimer's, and even IQ. It's no wonder everyone believes we're determined by our genes.

But if the working of genes is complex and subtle, as the research we've described shows, then something must be awry in the claims about finding genes "for" this or that trait. Geneticists Neil Risch and David Botstein wrote a commentary in *Nature Genetics* in 1996 describing the search for the gene for manic depression (Risch and Botstein 1996). They found that over the previous twelve years sixteen different research groups had announced the discovery of genetic linkages to manic depression (which translates in popular language into "gene for manic depression"). The problem—from a "one-gene, one trait" perspective—was that the purported gene had been correlated with fifteen different locations on eleven different chromosomes! Not lacking in humor, Risch and Botstein state that "one might argue that the recent history of genetic linkage studies for this disease is rivaled only by the course of the illness itself." They see the lack of consistency as an expression of the complexity of the illness on the one hand and not enough rigor in statistical analysis on the other. Evidently, the urge to find a genetic cause often overshadows the recognition of the complex nature of the phenomena.

As we have described elsewhere, even diseases that follow a more straightforward Mendelian pattern of inheritance, like sickle-cell anemia, are complex when looked at more carefully (Holdrege 1996). And it doesn't take much investigation to find that all of the characteristics or diseases listed above—none of which follow a Mendelian pattern—are strongly related to individual and environmental factors, as well as having some hereditary component.

The problem is that the isolation of a genetic factor is always based on a narrow theoretical and experimental framework. Or to put it in Kurt Goldstein's terms, genetics works with the method of isolation and therefore produces results that are valid only within that framework (Goldstein 1963). Take the example of amphetamine susceptibility. Scientists discovered that two different inbred strains of mice showed a very different relation to amphetamines: strain C mice preferred the box where it received injections of amphetamine, while strain D mice avoided this box (Cabib et al. 2000). You can already picture the headlines: "Scientists prove amphetamine addiction is hereditary." (How often we read such articles, only to discover that what we thought was a report about a human condition turns out to be an experiment with rats or fruit flies!)

But in this case the scientists were very careful and performed an additional experiment: they gave the mice less food over a period of time, while continuing amphetamine injections. Something unexpected occurred: strain D mice began to prefer the injection box, while the previously "addict-type" strain C mice avoided the box. A total reversal of the results! This example illustrates drastically what, in fact, is generally the case: a "fixed genetic predisposition" may actually be only one of many appearances (phenotypes) of an organism, and this particular appearance depends largely on the specific experimental and environmental circumstances under which the trait is observed.

Tinkering with Ourselves

The reduction of society to a community of believers in genetic determinism is, by itself, bad enough. But every worldview also has its practical effect on human action. The more we believe that genes determine our physical and mental constitution, the more we will be willing to tinker with those genes to change characteristics.

And this will occur in the name of human rationality. In 1998 a group of scientists met to discuss genetic manipulation of human beings, and the proceedings were published two years later (Stock and Campbell 2000). The participants promoted the view that science must progress and that genetic modification of human beings is inevitable: "Science proceeds and succeeds by doing. . . . What we're talking about here are incremental advances with enormous implications" (80). James Watson, the co-discoverer of the double helix model of DNA and the first head of the Human Genome Project, made the following comment:

> Some people are going to have to have some guts and try germline therapy without completely knowing that it's going to work. . . . And the other thing, because no one has the guts to say it, if we could make better human beings by knowing how to add genes, why shouldn't we do it? What's wrong with it? Who is telling us not to do it? I mean, it just seems obvious now. . . . If you could cure what I feel is a very serious disease—that is, stupidity—it would be a great thing for people who are otherwise going to be born seriously disadvantaged. We should be honest and say that we shouldn't just accept things that are incurable. I just think, "What would make someone else's life better?" And if we can help without too much risk, we've got to go ahead. (Stock and Campbell 2000, 79)

Watson is known for his blunt statements, revealing, we believe, a widespread sentiment that other scientists share but don't dare express: the path of genetic engineering leads to the human being, and we shouldn't close our eyes to this inevitable fact. The real challenge, in this view, is to convince the public. The book's editors, scientists Gregory

Stock and John Campbell, write: "To think rationally about ethical issues in germline engineering requires basic understanding of inquiry-based analysis and general scientific (biological) background. . . . If all scientists were to make a commitment to improving K-12 science education in their local communities, we might eventually have a society capable of thinking analytically and rationally about the challenges and opportunities of science—including germline engineering" (24).

In other words, people are not smart enough to see where science needs to take humanity. If we could get all elementary school children to isolate genes, middle school children to sequence them, and finally high school students to manipulate organisms with the genes, then perhaps we'd have the proper preparation. Of course, all learning about living organisms in their natural habitats would have to be dropped to provide space for such a high-tech curriculum. This would be the way to further "rational thinking."

In reality, what Stock and Campbell are aiming at is indoctrination in reductionism, so that people will lose the capacity to see through the weak and outlandish arguments of a Nobel laureate like James Watson. It's astounding that we've come so far that being rational is equated with tearing a narrow, genetic segment from the fabric of life and treating it as though it were everything. You're rational if you prohibit yourself from seeing how your sector of knowledge relates to a larger whole.

As we have shown, the results of modern genetics are shouting at us to wake up and see that we've got to start taking the whole organism seriously and to view genes in light of the organism and not only the other way around. Genetics began by defining genes in relation to a particular trait, ignoring the experimental and conceptual framework, and also ignoring the organism as a dynamic, changing entity. Now the emphasis should be on how an organism utilizes its genes within this broader context.

But the reductionist path is well worn and deeply entrenched. Once you're in it, it's hard to climb out. It's not easy to break out of habits and change an inner direction. It means giving up the security that comes with focusing on our own particular program that biases the mind from the outset. ("Understanding an organism means reducing its functions to underlying mechanisms.") Instead, our focus needs to be on entering the richness of the phenomena themselves and changing our viewpoints in order to do justice to what we discover. Instead of barraging the world with a monologue, we enter into conversation with it. How else can we hope to find deeper understanding and responsible ways of acting?

Chapter 8

Me and My Double Helixes

"What will you have done to your newborn," Bill McKibben (2003) asks, "when you have installed into the nucleus of every one of her billions of cells a purchased code that will pump out proteins designed to change her?" His answer is stark—and, we believe, misdirected:

> You will have robbed her of the last possible chance of understanding her life. Say she finds herself, at the age of sixteen, unaccountably happy. Is it her being happy—finding, perhaps, the boy she will first love—or is it the corporate product inserted within her when she was a small nest of cells, an artificial chromosome now causing her body to produce more serotonin? Don't think she won't wonder: at sixteen a sensitive soul questions everything. But perhaps you've "increased her intelligence"—and perhaps that's why she is questioning so hard. She won't even be sure whether the questions are hers. (47)

In his book *Enough: Staying Human in an Engineered Age*, McKibben repeatedly comes back to this point. A lover of running, he says that "if my parents had somehow altered my body so that I could run more quickly, that fact would have robbed running of precisely the meaning I draw from it"—the meaning that comes from exertions and achievements he could call his own (48). "If you've been designed and programmed to run, what meaning can running hold?" (55).

Likewise, noting that scientists "have pinpointed the regions of the parietal lobe that quiet down when Catholic nuns and Buddhist monks pray," he surmises that genetic engineers will before long be able to amplify the reaction.

> As a result, the minister's son may be even more pious than he is, but if he has any brain left to himself he will question that piety at the deepest level, wonder constantly whether it means anything or if it's so much brainwashing. And if he doesn't question it, if the gene transplant takes so deeply that he turns into an anchorite monk living deep in the desert, then his faith is utterly meaningless, far more meaningless than the one his medieval ancestors inherited by birthright. It would be a faith literally beyond questioning and hence no faith at all. He would be, for all intents and purposes, a robot. (48)

And so, too, there's the pianistically inclined mother who wants her child to be an even better pianist than she. But the point of piano playing lies in the meaning that is created through inclination and effort. "If the mother injects all that into her daughter's cells, she robs her daughter forever of the chance to make music her own authentic context—or to choose something else." The daughter would be a player piano as much as a human, "ever uncertain whether it is her skill and devotion or her catalogue proteins that move her fingers so nimbly" (48).

Too Much Liberation?

As happy as one might expect to be when a writer of Bill McKibben's stature draws attention to the troubling potentials of DNA manipulation, we fear that in this book he has unwittingly sided with the manipulators. But this will take a little explaining.

McKibben usefully sketches our progressive loss of human context, which is also a loss of meaning. The automobile wrenched us loose from local community; television isolated us from our immediate neighbors; divorce as a mass phenomenon cast a shadow of uncertainty over every family; and the natural world itself has been arbitrarily reshaped according to our habits and appetites, so that it no longer offers us "a doorway into a deeper world." But don't waste time asking whether these changes are good or bad, he advises us. They "came upon us like the weather," before we could do anything about them.

What, then, are we left with as a resource against meaninglessness? Only our individual selves. And that important truth brings McKibben to his punch line, which is that now, thanks to the genetic engineers, "we stand on the edge of disappearing even as individuals." Of course, the engineers put it in slightly different terms. They "promise to complete the process of liberation, to free us or, rather, our offspring from the limitations of our DNA, just as their predecessors freed us from the confines of the medieval worldview, or the local village, or the family."

But this, McKibben opines, "is one liberation too many": "We are snipping the very last weight holding us to the ground, and when it's gone we will float silently away into the vacuum of meaninglessness" (47).

Yet, unlike with those earlier challenges (although he does not explain why now and not then), we still have a choice when it comes to germline engineering. "What makes us unique is that we can restrain ourselves. We can decide not to do something that we are able to do. We can set limits on our desires. We can say, 'Enough'" (205).

We Are Not Our DNA

The problem is that McKibben's entire line of argument is self-defeating. "If you genetically alter your child and the programming works," he tells us, "then you will have turned your child into an automaton to one degree or another." As we heard above, the monk with genetically reinforced piety "would be, for all intents and purposes, a robot."

But if this is true—if we are, in this mechanistic sense, creatures of our DNA—then we are robots in any case. An entity that can be programmed is already an automaton. That's what it *means* to be an automaton. What difference does it make whether "chance events" programmed us, or someone in a lab coat? If, as McKibben insistently repeats, a twiddled bit of DNA substitutes for your meaningful self, then so, too, does an untwiddled bit of DNA.

McKibben is very good at showing how the engineers would treat our children as product lines, and would view the results of unsuccessful experiments as defective products. But he never offers a clear alternative to this product-view of ourselves. The parents who succumb to the lure of germline engineering would, he suggests, "be inserting genes that produced proteins that would make their child behave in certain ways throughout his life. You cannot rebel against the production of that protein" (58). Well, no more and no less can you rebel against the protein that would have been there without the engineering. Why, on his argument, are you not the product and slave of *that* protein?

McKibben reasonably asks us to exercise our freedom by saying

"Enough!" to the engineers. But there is startling dissonance in hearing someone argue for the urgency of free choice by asserting that our proteins determine our choices. By appearing to validate the scientist's (and the public's) conviction that we are our protein-producing DNA, McKibben is assisting the engineers' program. For while his commendable aim is to convince us to pull back from the eugenic brink, the fact is that those who think they are their DNA are exactly the ones who will clamor for a new and improved self, or at least for new and improved children.

Will the genetic engineers make our lives meaningless? This is ever so close to the truth, yet light years away from it. No one can, in absolute terms, rob someone else of meaning. What makes life meaningless is our rejection of meaning—a rejection we have already given expression to when we conceive ourselves as the product of DNA "mechanisms." The engineers are not making our lives meaningless; they are acting out the implications of their own flight from meaning by grasping whatever straws of pseudo-meaning they can find in their high-tech toys.

McKibben should have said, not that we are at risk of designing the individual self out of existence, but rather that we are directing unprecedented violence against it. We would make our offspring, in C. S. Lewis' phrase, the patients of our power (1965, 70), which is not at all the same as destroying their selfhood. Historically, the human individual has shown itself capable of surviving every imaginable insult, including those originating in the Gulag and Holocaust of the past century. It will also survive chemical assaults from the environment and chemical assaults from within its own body.

The real question today—a question McKibben's book only makes more poignant—is whether the individual can survive disbelief in his or her own existence.

A Concept on the Verge of Collapse

That the worshipers of machinery, efficiency, and power are engaged today in a fateful assault upon the human being is beyond all doubt. McKibben performs a valuable service by documenting this assault for a large audience from the mouths of the commandos carrying it out. There is no shortage of testimony. To take just two brief examples: Robert Haynes, president of the Sixteenth International Congress of Genetics, understands our ability to manipulate genes as indicating "the very deep extent to which we are biological machines." Likewise, Rodney Brooks of the MIT Artificial Intelligence Laboratory declares that interacting molecules are "all there is." They have produced the human body—"a machine that acts according to a set of specifiable rules. . . . We are machines, as are our spouses, our children, and our dogs." As for the contraptions that will surpass us, we should be under no illusions: "Resistance is futile" (quoted in McKibben 2003, 204).

McKibben makes as if to tackle this sort of gibberish. But in his eagerness to raise the alarm as shrilly as possible, he ends up granting far too much plausibility to the engineers. It is, after all, just laughable to claim that a particular gene, or any identifiable suite of genes, can account, in a coherent and manageable way, for intelligence or running ability or pianistic skill or piety. Certainly, as McKibben notes, there are people like the double helix celebrity James Watson who speak glibly of "going for perfection," as if this goal laid out an obvious course that could be traversed with reliable means. But one wonders whether Watson has ever devoted sixty seconds of his life to contemplating what it might mean for a human being to be perfect, or how we might get there.

The fact is that nearly all genetic engineers today have abandoned any simplistic "gene-for-this" idea. The idea has proven problematic enough when it comes to the most narrowly defined human diseases. Transpose it to deep character traits and skill sets and it is off the mark by what one can only call an astronomical order of magnitude.

Evelyn Fox Keller in *The Century of the Gene* (2000) provides an excellent review of the state of genetic research. Her long summary of the complications besetting the easy, simplistic notion of the gene culminates in this ironic observation: "At the very moment in which gene-talk has come to so powerfully dominate our biological discourse, the prowess of new analytical techniques in molecular biology and the sheer weight of the findings they have enabled have brought the concept of the gene to the verge of collapse" (69).

So it is that "the prospect of significant medical benefits—benefits that only a decade ago were expected to follow rapidly upon the heels of the new diagnostic techniques—recedes ever further into the future."

The Unity of the Organism

McKibben briefly alludes to these problems. It is, he grants, "unlikely that genes work quite as simply as the standard models insisted" (13).

This may sound comforting, he allows; "maybe there's not much to worry about; maybe it's a problem for the grandkids." But he quickly moves on: "In fact, however, all these qualifications mask the larger truth: *genes do matter.* A lot. That fact may not fit every ideology, but it does fit the data. Endless studies of twins raised separately make very clear that virtually any trait you can think of is, to some degree, linked to our genes. Intelligence? The most recent estimates show that half or more of the variability in human intelligence comes from heredity" (14).

Of course genes matter. All aspects of the human organism matter, and many of them are related to what we call "heredity," just as many of them are related to our cell membranes or to our hearts and would likely be affected, in some cases drastically, by an operation to modify the heart's functioning. Heredity affects everything, but then, too, the stuff and the interactions we (rather arbitrarily) group under the label of "heredity" are affected by all the rest of the organism. There is no way to slice up the organic unity that we are and say, without qualification, "This part unilaterally determines that." This sort of causality simply doesn't exist in the organism.

McKibben's argument for the effectiveness of genetic engineering has two steps. First, he cites evidence for the partial genetic determination of traits ranging from muscle mass to intelligence to homosexuality. Then, noting a history of accelerating technical success, he suggests that it's just a matter of time before we can reliably engineer these traits into our offspring.

This is a subtle mixture of truth and nonsense that desperately needs sorting out. Yes, we can be sure that more and more genetic "causes" for this and that will be found, much as we have been finding one substance after another that "causes" cancer. In fact, one can say with a great deal of confidence that a vast number of things, in some amount and via some possible pathway, can meaningfully be linked to cancer or its avoidance. That's just the way organisms work; one thing relates to another, however complex the pathways between. But when we begin discovering that "all kinds of things" are causally related to a particular condition, we also begin realizing that we haven't learned much about the condition at all—not, at least, so long as we remain stuck in a mechanistic, cause-and-effect mind-set, disregarding the governing unity and expressive tendencies of the organism as a whole.

There is no doubt at all that our technical capabilities will continue to develop at an ever-accelerating pace. It's possible that we will learn to stick this set of genes in that location with greater and greater precision. We certainly will continue to overcome previously "unsurpassable" technical barriers. And we will declare our efforts successful or unsuccessful in willful ignorance of all the ways the organism has shifted its entire structure and way of being in response to the unasked-for invasion. It will be enough, for the technician, that some desired effect was noted. And, yes, McKibben is right: some of these effects will be commercially valuable. We can expect to see much of the technological sickness he describes.

But none of this represents success at the kind of global re-engineering of the organism and the individual that McKibben envisions. Certainly we can pursue such engineering in a negative sense, perhaps all the way to making the physical body humanly uninhabitable. We can throw up decisive genetic obstacles to the individual's self-expression through his own body—we could, to be trivial, intentionally or unintentionally disable critical parts of the nervous system.

We will doubtless also be able to make some collection of changes known to have a bearing on, say, intelligence. But this is very different from a knowledgeable, systematic, or coherent redesign of the organism, which would require a different kind of science from what we now have. And it doesn't address the question of the individual self at all. If we pursue this path, we may arbitrarily interfere in the destinies of our fellows in countless novel ways, and we may count many isolated alterations as "improvements," but we will not be engineering superior human beings.

Intertwined Lives

McKibben is emphatically right in his central contention: the gene manipulators are promising us nothing less than disaster. But our only hope for avoiding the disaster lies in an ability to move beyond the current terms of the debate.

McKibben, as we saw, believes that if his parents had altered his body to make him a faster runner, it would have robbed his running of its meaning. But—as he notes without adequately exploring the fact—all parents do have the power to alter their children's bodies, and they always exercise that power. When we leave aside the fact that by pairing off as they did they determined a great deal about their offspring's genetic heritage, there remain all the effects of upbringing.

The quality of a child's diet, for example, can make the difference between a superb runner and an obese nonrunner. The kind of activity—or inactivity—the parents encourage while young bones, muscles, and nerves are developing sets bounds to what the adult may eventually achieve. And parental carelessness—or, worse, downright abuse—may result in an injury radically limiting the child's potentials as a runner.

The young child can hardly be expected to override or control all such parental influences. Moreover, these influences extend beyond the body, into the innermost regions of the psyche. Whether a child is brimming with confidence, ready to take on every new challenge as a runner, or instead shies away from such challenges may depend in part upon the parents' love and supportiveness.

The moral is simple: we are caught up in each other's destinies. There is no escaping the fact. This, however, is not to dismiss the importance of our interactions with each other as "merely routine." Quite the opposite. One way to put it is to say that McKibben's concern for the grave implications of genetic engineering should be extended to all those other ways in which we "engineer" one another's destiny. We may have taken these far too lightly. We do not, alone, become what we are.

But nothing in this line of thought justifies our arguing that, because DNA manipulation recklessly affects someone's destiny, this manipulation therefore reconstitutes the self or robs it of meaning. So far as we can tell, McKibben does not offer a single sentence in justification of this primary contention. A mere assumption—and a pernicious one at that—is the ruling center of his argument. Furthermore, the clear (if unspoken) implication of his argument is that if your physical body and its chemistry *have* been dramatically and irreversibly shaped by an abusive parent or anyone else, then your life, too, must be to that extent meaningless.

"Germline engineering destroys the meaningful existence of the individual." It is hard to give up such a clear-cut line of defense when one feels under the shadow of an extreme danger. But surely our long-term hope hinges on truly outlining the danger rather than misrepresenting it.

Self and World

Any such assessment, it seems to us, must begin with this truth: the skin of our bodies does not constitute the boundary between self and world. Our physical bodies belong to the outer world, even though we obviously have a special relationship with this particular part of the world. And just as a "blow of destiny" from the world—say, a freak, disabling injury to your body—does not subvert that core place within you where you experience yourself as a free and spiritual being, neither do the assaults of the engineers upon your bodily DNA destroy this inner reality.

As Craig Holdrege makes clear in *Genetics and the Manipulation of Life,* the organism treats a bit of injected DNA much as it treats alien elements introduced into its external environment: it adapts in its own distinctive manner, an adaptation expressing its inner way of being (which is exactly what makes the organism-wide results of genetic engineering so unpredictable by an engineering mentality). Injuries, whether inflicted accidentally or by engineers, may severely limit our possibilities of expression in the world—and may do so in morally reprehensible ways—but this is not the same as destroying the self that responds freely and in its own way to these limitations.

McKibben describes the death of his childhood friend, Kathy, from cystic fibrosis. Pondering whether he would opt for germline engineering to overcome such a disease, he worries about the slippery slope this would put us on. The problem, he says, is that there is no clear line between repair of obvious disorders and the sort of enhancement that redefines the self: "Soon you're headed toward the world where Kathy's lungs work fine, but where her goodness, her kindness, don't mean what they did. Where someone's souping up her brains or regulating her temper, not just clearing up her mucus."

There *is* a slippery slope, and we *should* avoid it. But we can't choose to do so by losing sight of the individual who must do the choosing—the individual who, for McKibben, seems always about to vanish into a set of controlling mechanisms.

Kathy's kindness could *never* mean in one situation what it meant in another. But this does not suggest the absence of her self in either case. We all know that it's much easier to be kind in social contexts where this is encouraged and supported from all sides. It's much harder in an every-man-for-himself environment. Kindness while under the influence of Prozac and kindness while undergoing chemotherapy may express quite different potentials of one's personality. But either there's a self capable of manifesting itself through all the differing physical and social conditions of the individual's life, or there isn't. And the conviction that there is—a conviction that must underlie any solution to our current quandary—is one that McKibben seems unable to muster with sufficient force to overcome the mind-set of the engineer.

The self-doubt that McKibben ascribes to the patients of the engineer's power ("Is this really *my* character?") is self-doubt we ought to feel in any case, since we are always tempted to abdicate the self's true achievement by yielding to mechanism. Do your responses at this moment genuinely reflect the freedom of your self, or are they rather indicative of the good meal you just enjoyed—or of the powerful hunger now voicing itself in your blood chemistry?

Or, if you are born with a temperamental inclination toward equanimity, you might well ask: how much of your deepest potential have you yet brought to birth through this physically mediated temperament? What new aspects of yourself would you discover if you ventured into extreme circumstances, forcing yourself out of the comfortable, rather too-easy pattern of your life? Do you avoid new challenges precisely in order to preserve your highly valued equanimity?

Such questions about the conditioning factors of one's life do not testify to the loss of self; they are a deep expression of the self. They exemplify its power to transcend all material conditions. To question something is already to have separated one's essential identity from the thing. McKibben has it backward when he equates the physical body (DNA) with the self and deprecates the self's questioning of itself as lostness. His stance serves the purpose of the engineers perfectly. Discount the self that can question and rise above the material conditions of life, and all you have left is a mechanism fit for tinkering.

Freedom and Limitation

If it's true that we unavoidably affect each other's destinies—for ill, but also for good—then everything hinges upon our understanding of this mutuality. And the first thing to grasp is that healthy human exchange is, and is essentially, a matter of mutuality. We are called to engage each other in a mutually respectful dance or conversation, which is very different from unilateral manipulation. Conversation or manipulation: *this* is the decisive distinction.

Two people in conversation meet and accommodate to each other. Each gives and each receives—the giving is prerequisite for the receiving and the receiving for the giving. The exchange is irreducibly moral, as is every meeting between self and other. One cannot talk about the good-for-me except in relation to the good-for-others. No human being grows and develops in the sense that counts most deeply except by helping others to grow and develop.

Such mutuality extends even to the relation between parent and tiniest child. Martha Beck's story in *Expecting Adam* (1999) reminds us in a startling way that a child, even a yet-unborn child, can speak powerfully in its own behalf, summoning from its environment the crucial elements of its destiny.

But we don't need such an extraordinary story in order to see the truth. Every attentive mother and father knows that their child, though lacking powers of intellectual articulation, has yet many voices for expressing its distinctive character and needs. Conscientious parents do not find themselves unilaterally determining the shape of their child's life; they are forever *responding* to what comes to meet them, which as often as not is unexpected. They struggle to make room for the unforeseen potentials the child is ceaselessly declaring.

This kind of mutuality characterizes all worthwhile human interaction. One result of this is that the Other, whose needs you must bear even as he bears yours, becomes not only essential to the realization of your destiny—a gift to you—but also a kind of burden, limiting your freedom. But the gift and the limitation are thoroughly intertwined, so that one turns out to be the face of the other.

In general, limitation and suffering, which we so often inflict upon one another, are inseparable from the highest gifts we receive. We must always work to overcome limitation and reduce suffering in the world, but if in this work we remain blind to their necessary and positive role, our work will be destructive.

Freedom is empty without the necessities that bind us. If we were able to act with complete, arbitrary abandon to achieve anything we wanted without restriction, it would mean we were able to do nothing significant. It would mean there was no constraining lawfulness, no order and regularity in the world, in which case our activity could have no coherent or meaningful effects. Without limitation and necessity (and the suffering they bring with them), there is no freedom.

Acting in Ignorance

Every philosopher who has ever looked at the problem of freedom and necessity has recognized this interweaving of the two. But if one or two of the vocal, best-selling advocates of eugenic engineering have ever attempted some such reflection, we haven't seen evidence of it. Those who speak of removing human limitation and going for perfection don't seem to have the slightest clue about the inseparability of meaningful achievement and limitation. And they appear perfectly content to talk about altering the physical "machinery" of the individual without any consideration of the mutuality essential to nontotalitarian human exchange.

What makes this infinitely worse is that, when they enter the laboratory, they don't know what they're doing—a point McKibben unfortunately obscures with his unrealistic depiction of the state of the art. The genomic engineer is carrying out his manipulations in utter ignorance of their broad, ramifying, and untraceable effects, and is either unaware of his ignorance or, much more likely, frighteningly casual about it.

The bedrock principle of the organism is that everything is connected to everything else—and in ways we have scarcely begun to understand. To alter the human genome today through the engineer's techniques is the moral equivalent of flipping a coin to determine whether a child will be educated in a public school, on the streets, at home, or in a prison for hardened felons. The difference, of course, is that there is at least a well-defined set of alternatives in mind when the coin is flipped; the genetic engineer who plays the DNA roulette wheel cannot begin to conceive the range of unknown but possible effects of this action—effects that may continue on down the germ line through countless generations.

Perhaps you will ask, "How *can* an element of mutuality enter into the engineer's dealings with the yet-unborn—or with the yet-unconceived?" If this seems to you an obvious impossibility, then that itself places a sobering question mark over these dealings. But if you are among those who see the citadel of materialism and mechanism being weakened and undermined from all sides, you may suspect that a mutual exchange between a parent and an incarnating child is not altogether unthinkable. In this case you cannot rule out, in absolute terms, all future application of genetic engineering techniques to the unborn—assuming, of course, that we eventually get the kind of qualitative science that would make such techniques meaningful and reliable in their implications for the entire organism.

A Bifurcation of Humanity

A final note regarding the Lee Silvers, Hans Moravecs, and Ray Kurzweils (all leading characters in McKibben's book) and their tales of the twilight of the human race as we have known it. These would-be prophets of a post-human future have found a rewarding niche for themselves proclaiming in the most outrageously satisfied manner they can contrive, "The end is near!" We can learn a great deal from them about certain tendencies of the technological mind-set, but not much at all about human freedom, the self, or truth, beauty, and goodness. To allow their rhetoric to determine the form of the discussion when you are concerned with responsible assessment of human nature and the shape of the future is to give up all clarity of thought at the very outset.

The "prophets," however, are justified in feeling a certain revulsion when they hear of absolute limits to human development, of challenges refused, of human achievement that has come far enough. They are correct: we will never have come far enough. Our dual responsibility is to accept our limits *and* to work against (or rise above) them in the knowledge that no limits are absolute, just as no freedom is absolute. Our life is our growth and development—growth and development within a context that forever limits, disciplines, and shapes us even as we forever reshape and transcend that context.

What the prognosticators miss is the crucial truth: in the end, all enduring achievements—the only ones we can ever be satisfied with—are inner ones. They are achievements of the spirit. The effort to conceive what we want in terms of outward mechanisms (which include the body's "mechanisms") is not a stretching toward new horizons, but a darkening of those horizons. The only truth in all the frenzy of post-human prediction is that we *can,* through inner abdication, bring about a twilight of the race. We are being urged toward this goal from all sides.

But perhaps "regression" is more accurate than "twilight." You can't read the futuristic scenarios and personal hopes of the re-engineers of humanity without being struck by the utter childishness of it all. Genetic modifications that will save us from the necessity of bodily excretion; nano-contrived plants that look exactly like orchids but can grow in frigid climes; robots that wait on us like slaves; a cyber-nano-genetically engineered "elite race of people who are smart, agile, and disease-resistant"; nanobot swarms able to wander the human bloodstream and keep us eternally healthy; technological horns of plenty that will convert every "desolate" village into "a Garden of Eden, with widescreen TVs and cappuccino machines for all," . . . and so on ad infinitum.

And many of these visions come from the same people who delight in ridiculing the "childish hopes" of the traditionally religious!

We may well be headed toward the kind of two-class society many of the engineers envision. And we can rightly fear that the lower, "unenhanced" class will be left pitifully behind. But this underclass will consist of the infantilized portion of the race—a group of people so mesmerized by what they see as the promise of technology that they will give up their own development. In their arrested, technology-fixated state, with their lightspeed tools of calculation and well-honed manipulative skills (wonderful pacifiers of the human spirit) they may, for who knows how long, wield the external power in society. But it will be the power of the child-tyrant. Meanwhile, a wiser humanity will continue maturing those inner powers of imaginative insight and moral resolve that just may, in the end, enable them to save the tyrants from themselves.

We wish we could say that Bill McKibben's opus is likely to encourage this wiser movement toward an ever-deepening human future. But a book that sows self-doubt in the face of the technological assault does not promise much encouragement.

Chapter 9

Logic, DNA, and Poetry

In January 1956, Herbert Simon, who would later win the Nobel Prize in economics, walked into his classroom at Carnegie Institute of Technology and announced, "Over Christmas Allen Newell and I invented a thinking machine." His invention was the "Logic Theorist," a computer program designed to work through and prove logical theorems. Simon's casual announcement—which, had it been true, would surely have rivaled in importance the Promethean discovery of fire—galvanized researchers in the discipline that would soon become known as artificial intelligence (AI). The following year Simon spoke of the discipline's promise this way: "It is not my aim to surprise or shock you. . . . But the simplest way I can summarize is to say that there are now in the world machines that think, that learn and that create. Moreover, their ability to do these things is going to increase rapidly until—in a visible future—the range of problems they can handle will be coextensive with the range to which the human mind has been applied" (Simon and Newell 1958).

There was good reason for the mention of surprise. Simon and his colleagues were, in dramatic fashion, surfing the shock waves produced by the realization that computers can be made to do much more than merely crunch numbers; they can also manipulate symbols—for example, words—according to rules of logic. The swiftness with which such programmed logical activity was equated, in the minds of researchers, to a humanlike capacity for speech and thought was stunning. And, during an extended period of apparently rapid progress, their faith in this equation seemed justified. In 1965 Simon predicted that "machines will be capable, within twenty years, of doing any work that a man can do" (Simon 1965, 96). MIT computer scientist Marvin Minsky assured a *Life* magazine reporter in 1970 that "in from three to eight years we'll have a machine with the general intelligence of an average human being . . . a machine that will be able to read Shakespeare and grease a car."

The story is well told by now how the cocksure dreams of AI researchers crashed during the subsequent years—crashed above all against the solid rock of common sense. Computers could outstrip any philosopher or mathematician in marching mechanically through a programmed set of logical maneuvers, but this was only because philosophers and mathematicians—and the smallest child—were too smart for their intelligence to be invested in such maneuvers. The same goes for a dog. "It is much easier," observed AI pioneer Terry Winograd, "to write a program to carry out abstruse formal operations than to capture the common sense of a dog" (Winograd and Flores 1986, 98).

A dog knows, through whatever passes for its own sort of common sense, that it cannot leap over a house in order to reach its master. It presumably knows this as the directly given meaning of houses and leaps—a meaning it experiences all the way down into its muscles and bones. As for you and me, we know, perhaps without ever having thought about it, that a person cannot be in two places at once. We know (to extract a few examples from the literature of cognitive science) that there is no football stadium on the train to Seattle, that giraffes do not wear hats and underwear, and that a book can aid us in propping up a slide projector when the image is too low, whereas a sirloin steak probably isn't appropriate.

We could, of course, record any of these facts in a computer. The impossibility arises when we consider how to record and make accessible the entire, unsurveyable, and ill-defined body of common sense. We know all these things, not because our "random access memory" contains separate, atomic propositions bearing witness to every commonsensical fact (their number would be infinite), and not because we have ever stopped to deduce the truth from a few, more general propositions (an adequate collection of such propositions isn't possible even in principle). Our knowledge does not present itself in discrete, logically well-behaved chunks, nor is it contained within a neat deductive system.

It is no surprise, then, that the contextual coherence of things—how things hold together in fluid, immediately accessible, interpenetrating patterns of significance rather than in precisely framed logical relationships—remains to this day the defining problem for AI. It is the problem of meaning.

DNA's Ever-Receding Secrets

On February 28, 1953, Francis Crick and James Watson burst into the Eagle pub in Cambridge, England, where (as Watson later recalled) Crick spilled the news that "we had found the secret of life." The secret, as the world now knows, lay in the double helical structure of DNA. Looking back on Crick and Watson's revelation fifty years later, the editors of *Time* would refer to "the Promethean power unleashed that day."

It was, however, slightly strange for Crick and Watson to announce the revelation of a secret that came in the form of a code they did not understand and a text they did not possess. Yet the double helix, by all accounts, came in just that way. This is why we have been treated, during the intervening fifty years, to the celebration of one code-breaking and text-reading victory after another, culminating most recently in the Human Genome Project. Only now, we're told, has the full text of the deciphered Book of Life been laid out before the eager eyes of genetic engineers.

The celebration—and also the expense—of this latest victory has been unparalleled in the history of biology. So, too, has the orgy of self-congratulation and utopian prediction. The completion of the genome project, many scientists declared, would quickly enable us to slay the demons of genetically linked disease, after which we would employ designer genes to create an enhanced race of superhumans. The giddiness reached its frothy zenith when Nobel laureate and molecular biologist Walter Gilbert observed that we will pocket a CD carrying the code for our personal genomes and say, "Here's a human being; it's me!" (1992, 96).

But wait! Hold off on the celebration. Now it appears there's one small remaining obstacle on the path to unprecedented self-knowledge. Yes, we have discovered the alphabetic text of the Book of Life, but it turns out we still can't actually read it. For this, according to the current story, we need a new project—one that will dwarf even the human genome effort. We must unravel the functioning of the body's 100,000 or more proteins—molecules so deeply implicated in every aspect of the organism (including its genetic aspects) that the attempt to understand them looks suspiciously like the entire task we began with: to understand life.

The secret of life, it appears, is wrapped within layer after layer of mystery, each one requiring its own decoding, and each one extending further through the biochemistry of the whole organism. Where, then, is the single, controlling secret? If by their own admission they still cannot read the DNA text of the Book of Life, how can molecular biologists pronounce so confidently on the nature and absolute importance of its meaning? And if they can achieve the reading only through recourse to everything else going on in the organism—that is, if they must in effect read the whole organism—then how can they know that the entire secret resides in one, small, still mostly undeciphered portion of the overall text?

Does Logic Make a Text?

Clearly there is a profound faith at work here. It is, in fact, the same faith that motivated Herbert Simon and his fellow AI researchers: once they laid hold of an apparently mechanizable logic, they just couldn't help themselves. The mechanism and logic *must* explain everything else! That is how they expected a mechanically conceived world to work, whether they were dealing with human speech and thought or the genetic text of the Book of Life.

What many molecular biologists seem to have been longing for, like most scientists working within a mechanistic tradition, was a world of life and thought driven by a neatly controlling syntax that played itself out with something like cause-and-effect necessity. They imagined this causal necessity much as they imagined the external impact of particle upon particle, molecule upon molecule, where one well-defined thing "makes" another happen.

And if this is how things work, then why should they have worried about what the Book of Life would turn out to say whenever they managed to read it? Their confidence that they had wrested the textual secret of life from the cell's nucleus even before they had a clue to its reading is the proof that they were not really thinking in textual terms. It wasn't the still-unknown meaning of the text that excited them so much as their conviction that a code—a cut-and-dried, mechanizable logic—had been found for preserving certain "machine states" from one generation to the next. Surely, they thought, the discovery of such a mechanism—seductive and unqualified in its clarity and reassuringly necessary in the attractions and repulsions of its logical atoms—would explain everything.

It was in this spirit that the so-called Central Dogma of Molecular Biology gained hold in the 1960s. According to this deeply influential doctrine, genetic information flows in one direction only, from genes to proteins. As science historian Evelyn Fox Keller paraphrased the Central Dogma: "DNA makes RNA, RNA makes protein, and proteins make us" (2000, 54). The doctrine, as commonly interpreted, with its genetic determinism and command-and-control view of DNA, paved the simplest, most direct highway to a mechanistic understanding of the organism.

Putting Genes in Context

But the highway proved to be little more than a long, rutted detour. The straightforward, neatly determining logical structure envisioned by geneticists—a structure that beckoned like a Holy Grail during the Human Genome Project—has progressively transformed itself into a seething cauldron of endlessly complex dynamic processes extending throughout the organism. The crucial problem for genetic determinism and the once-prevailing Central Dogma is that biochemical cause and effect within the cell, as in the organism as a whole, never proceeds in one direction alone. There is an irreducibly complex, organic web of causation operating throughout the entire cell and organism.

The string of discoveries supporting this conclusion is not contested. We now know that one gene can produce many different proteins, depending on complex processes that are orchestrated not only by DNA, but also by proteins themselves. Moreover, one protein is not necessarily one protein. For example, depending on the presence of so-called chaperon proteins, a given chain of amino acids (the constituent elements of protein) may fold in different ways. These various foldings in turn shape the overall structure and functioning of cell and organism.

The supposedly linear structure of letters, words, and sentences into which DNA has been decoded simply does not articulate a clean, unambiguous, command-and-control authority sitting atop a hierarchical chain of command. Only a misguided preoccupation with an imagined set of idealized syntactical relationships could have led researchers to dismiss the greater part of DNA—nearly all of it, actually—as "junk DNA." The junk didn't seem to participate in the neat controlling sequences researchers were focused on, and so it seemed irrelevant. But more recently the erstwhile junk has been recognized as part of a "complex system of distributed regulation" in which "the spacing, the positioning, the separations and the proximities of different elements . . . appear to be of the essence" (Moss 2003, 191).

But even more devastating for the centralized command-and-control view has been the discovery of "epigenetic" processes. These yield hereditary changes that are not associated with structural changes in DNA at all. Rather, they arise from alterations in how the rest of the organism marks and employs its DNA. And beyond this, researchers have been exploring effects upon DNA from the larger environment. In a dramatic reversal of traditional doctrine, investigations of bacteria show that gene mutations can arise from—can even be guided by—environmental conditions in a nonrandom way (see chapter 5). In sum, genes are no more the self-determining cause of everything else in the organism than they are themselves the result of everything else.

Finally, we have seen a startling demotion of the human genome in size relative to other organisms. The most recent and near-final estimate by the Human Genome Project puts humans in possession of 20,000 to 25,000 genes—this compared to at least 25,000 for a tiny, primitive, semi-transparent worm, *Caenorhabditis elegans.* If genes constitute the one-way controlling logic or master program determining the potentials of the organism, then finding such unexpected gene counts is rather like discovering we could implement all the programs of the Microsoft Office suite using only the minuscule amount of program logic required for a simple daily greeting program.

Reviewing the history of misdirection surrounding the gene, cell biologist Lenny Moss writes,

> Once upon a time it was believed that something called "genes" were integral units, that each specified a piece of a phenotype [that is, a trait], that the phenotype as a whole was the result of the sum of these units, and that evolutionary change was the result of new genes created by random mutation and differential survival. Once upon a time it was believed that the chromosomal location of genes was irrelevant, that DNA was the citadel of stability, that DNA which didn't code for proteins was biological "junk," and that coding DNA included, as it were, its own instructions for use. Once upon a time it would have stood to reason that the complexity of an organism would be proportional to the number of its unique genetic units. (Moss 2003, 185)

Today, as we have already heard from science historian Evelyn Fox Keller in the previous chapter, the findings of the past few decades "have brought the concept of the gene to the verge of collapse." In fact, she adds, "it seems evident that the primacy of the gene as the core explanatory concept of biological structure and function is more a feature of the twentieth century than it will be of the twenty-first" (Keller 2000, 9, 69).

Taking Our Words Seriously

To point out the inadequacy of the Central Dogma will strike most geneticists today as anachronistic. "We long ago quit believing such a simplistic doctrine." And, in fact, you will find almost everyone regularly disclaiming the "gene-for" view—that is, the belief that for many or most traits of the organism there is a gene, or a few genes, that account for the trait. "We know it's much more complicated than that"—so the disclaimer runs. In the face of such protestations, recital of the history of misdirection begins to seem unfair. After all, scientists must be allowed the freedom to speculate, as long as they keep learning along the way. What's important is the knowledge they eventually arrive at.

But does the painfully repetitive history of genetics and AI suggest that they have in fact been growing beyond their infatuation with logical mechanisms? The best way to answer this question is to elucidate the central misdirection in the history under discussion.

The real significance of the overheated rhetoric of the Human Genome Project lies in the seemingly unstoppable appeal by molecular biologists to language and thought—that is, to book, word, letter, code, translation, transcription, message, signal, and all the rest. Or, to employ the most universal term today: information. This resort to a terminology so brazenly *mental* in origin appears to be a stunning reversal. Just a few decades ago we still lived within the long historical era during which it was unpardonable for the natural scientist to draw his explanatory terms from intelligent activity. What changed?

Crucially, the age of cybernetics and computation arrived. This brought with it, for many researchers, the promise of the mechanization of language and thought. Suddenly it became respectable to invoke human mentality in scientific explanation because everyone knew you weren't really talking about mentality at all—certainly not about anything remotely resembling our actual mental experience. You were invoking computational mechanisms. So the change was less a matter of assigning human intelligence to the mechanically conceived world than of reconceiving human intelligence itself as mechanical performance.

Of course, we have seen that the equation of mechanical computation with mentality was based on the extraordinarily naive assumption that machine logic is the essence of thinking and language. But looking past this reductionism, we find that molecular biologists have glimpsed something extraordinarily deep. If they have been driven to textual metaphors with such compelling, seemingly inescapable force, it is because these metaphors capture a truth of the matter. The creative processes within the organism are *word-like* processes. Something does speak through every part of the organism—and certainly through DNA along with all the rest. Molecular biologists are at least vaguely aware of this speaking—and of the unity of being it implies—and therefore they naturally resort to explanations that seem to invoke a *being who speaks.*

The problem is that insistence upon textual *mechanisms* blinds one even to the most obvious aspects of language—aspects that prove crucial for understanding the organism. If someone is speaking to you in a logically or grammatically proper fashion, then you can safely predict that his next sentence will respect the rules of logic and grammar. But this does not even come close to telling you what he will say. Really, it's not a hard truth to see: neither grammatical nor logical rules determine the speech in which they are found. Rather, they tell us only something about *how* we speak. (See also chapter 13.)

If molecular biologists would reckon fully with this one central truth, it would transform their discipline. They would no longer imagine that they could read the significance of the genetic text merely by laying bare the rules of a molecular syntax. And they would quickly realize other characteristics of the textual language they incessantly appeal to—for example, that meaning flows from the context into the words, altering the significance of the words. This is something you experience every time you find yourself able, while hearing a sentence, to select between words that sound alike but have different meanings. The context tells you which one makes sense and, more generally, it can modify the meanings of *all* the words.

The role of context is pervasive. As poets know very well, even the word "prophet" in the two phrases "old prophets" and "prophets old" carries different ranges of meaning (Barfield 1973, 41). If DNA is really like a text, then plasticity of the gene must be one of the rock-bottom, fundamental principles of heredity.

Conversation and Poetry

There is no need for geneticists to endure lectures from philologists, however. As we have seen, all this is exactly what their own discoveries of the past fifty years have been shouting at us. In the ongoing conversation between word and text, part and whole—and contrary to the command-and-control view—we find the context of the organism *informing* the genetic text at least as much as the genes can be said to inform the organism. This is the underlying truth that science historian Lily Kay trades on when she writes: "once the genetic, cellular, organismic, and environmental complexities of DNA's context-dependence are taken into account," we might find that genetic messages "read less like an instruction manual and more like poetry, in all their exquisite polysemy [multiplicity of meaning], ambiguity, and biological nuances" (2000, xviii–xix).

What this means practically is that "we gain a knowledge of genes . . . only through knowledge of the organism as a whole. The more knowledge we have of the organism as a whole, the more information we have. *This information is not in the genes; it is the conceptual thread that weaves together the various details into a meaningful whole*" (Holdrege 1996, 80, emphasis in original).

The weaving together is a *conversation,* not a merely mechanical unrolling of a logically compelling sequence. When we speak of such things as *messenger* RNA, the conversational context should be obvious. It makes no sense—or, at least, no sense that biologists have yet explained—to speak of a message without a recipient capable of a certain understanding, and without a context for determining how the message is to be construed. If we eliminate these things from the picture, we have a message without meaning, which is no message at all. The question, then, is whether the researchers really believe their own terminology.

They ought to. Everything we have been learning about the genome points to the significance of its conversational context. As Lenny Moss puts it: "If the sum total of coding sequences in the genome be a script, then it is a script that has become wizened and perhaps banal. It wouldn't be the script that continued to make life interesting but rather the ongoing and widespread *conversations about it*" within the biochemistry of the organism (2003, 190, emphasis in original).

Actually, it is not so much the script that is banal as the reduced, syntactic reading of it. As Moss himself reminds us, the script is a dynamic one, subject to continual and rapid changes with profound significance—"transpositions, amplifications, recombinations, and the like, as well as modulation by direct chemical modification" (110). There is a lively conversation going on here, but it is one in which our genes are caught up, not one they are single-handedly dictating.

Words of Explanation

Language is the very soul and substance of explanation itself. The reason for this can only be that the world we are explaining has something language-like about it. When we offer a scientific explanation for some aspect of the world, we necessarily assume that the meaning of our words is at the same time the meaning of the chosen aspect of the world. If this so-called intentionality of language—its being about something else and not just about itself—were not born of the world's word-like character, then our scientific explanations could tell us nothing about reality. The world must in some sense be a text waiting to be deciphered. This is why the scientist can, in fact, decipher it into the text of a scientific description.

So in reality *all* scientific explanation is founded upon an appeal to the word. The irony lies in the fact that precisely where the computer scientist and geneticist resort explicitly to "word" and "text," what we actually see is a concerted attempt to substitute wordless logic and computational mechanisms for language.

Most fundamentally, this stance takes the form of an attempt to explain words themselves as if they were objects. No longer understanding words as the light bearers through which the things of the world become visible and meaningful, we instead take these words merely as additional things requiring explanation. That is, we want to understand our explanatory words as if they themselves were nothing more than causal results of processes going on in the world they explain. There is something gravely misconceived in this effort to explain explanation itself—and all the more when the effort involves an appeal to mechanisms stripped as far as possible of their word-like (and therefore of their explaining) nature. It is rather like trying to *prove* the validity of logic—or, in other words, trying to prove the validity of the instruments of proof—and to do so by invoking physical laws. A fool's task.

And yet this was the path computer scientists went down when they came to believe they could explain speech (and thought) as manifestations of computational devices. Their aim was to explain our powers of explanation by appealing to something not having the essential character of explanation. The result could only be nonsense, which is why the researchers quickly began arbitrarily projecting language back into their wordless explanatory devices.

At its worst, the projecting became extraordinarily crude. All one needed to do was to label programs and data structures with terms like UNDERSTAND and GOAL, and then mindlessly assume that the programs actually had something to do with understanding or goal-seeking. Such nonsense eventually became downright embarrassing. In 1981 computer scientist Drew McDermott published an essay entitled "Artificial Intelligence Meets Natural Stupidity" in which he ridiculed the use of "wishful mnemonics." He wondered aloud whether, if programmers used labels such as G0034 instead of UNDERSTAND, they would be equally impressed with their clever creations.

Likewise, McDermott commented on Herbert Simon's "GPS" program, written as a much more ambitious successor to the Logic Theorist: "By now, 'GPS' is a colorless term denoting a particularly stupid program to solve puzzles. But it originally meant 'General Problem Solver,' which caused everybody a lot of needless excitement and distraction." He went on to say, "As AI progresses (at least in terms of money spent), this malady gets worse. We have lived so long with the conviction that robots are possible, even just around the corner, that we can't help hastening their arrival with magic incantations. Winograd . . . explored some of the complexity of language in sophisticated detail; and now everyone takes 'natural-language interfaces' for granted, though none has been written. Charniak . . . pointed out some approaches to understanding stories, and now the OWL interpreter includes a 'story-understanding module.' (And, God help us, a top-level 'ego loop.')" (McDermott 1981, 145–46).

The geneticist's strategy with genes was much like the computer scientist's strategy with program modules. All that was needed was to put a label on the gene associating it with such-and-such a trait and—presto!—the gene now spoke a meaningful language. The only problem is that these neatly labeled genes are forever disappearing as rapidly as they are discovered—or, rather, they lose their neat, causal identity against a background of extraordinary complexity. What stands on the biochemical and supposedly causal side of the relation never clearly relates to the trait, and certainly fails to explain it in any adequate sense.

This is because the trait—whether it is dark skin, green eyes, cancer, or an aggressive tendency—is quite properly understood in qualitative and meaningful (word-like) terms, whereas the "causal" gene remains at the level of mechanism, not language. Causes and mechanisms are no more able to originate meanings than grammatical rules are able to originate or explain the things we say (Talbott 1995).

There could hardly be a surer indication of the insecure and disturbed foundations of science than we find in all the confusion over word and text—a confusion that can lead only to the destruction of science as a discipline of understanding rather than merely of effective technique. How much confidence can we place in the understandings conveyed through an enterprise whose verbal and conceptual instruments of understanding are so badly damaged? If there is to be a scientific Prometheus for our day, he must bring the fire of meaning into our various theoretical languages—languages that, in their current, desiccated state, are like dry tinder eager for the blaze. And it is almost as if geneticists, with their ceaseless invocation of word and text, have been unconsciously calling down the tongues of flame.

Such a conflagration will doubtless consume a great deal. But it may also purify and transform. If the concept of the gene really has been brought to the verge of collapse, we can hope that in our revitalized understanding the gene will truly speak with all the creative and clarifying power of the word—because the entire organism speaks through it. Then its language of wholeness will belong as much to the poet as to the scientist, and we will hear within its rhythms and cadences a song of destiny in which we ourselves are singers.

Part III

To Be an Organism

Chapter 10

The Cow

Organism or Bioreactor?

The tall tree on the next page probably does not match your mental picture of a typical white oak (*Quercus alba*). The trunk appears disproportionately long and narrow and the crown is small compared to the trunk. Is this tree unhealthy? No. It is perfectly healthy, but it is shown out of context.

Set in the middle of a forest, it would look like most of the other trees in a bottomland deciduous woodland in northeastern North America. You have to imagine this single tree surrounded on all sides by other trees of similar height with long trunks reaching skyward—mainly sugar maples, red maples, and red oaks. They have grown in concert with one another, perhaps out of an abandoned pasture, for about eighty to a hundred years. As they spread their crowns, they produced shade for one another, and the dominant growth direction was upward into the light-filled space above. The lower branches, which never grew to great size, died off in the increasingly shady environment of the upward-shooting trees. In this way, the long, branchless trunk developed, and we need to imagine the seemingly meager crown of the individual trees as part of the larger green canopy of the whole forest.

The shorter tree is also a white oak, with a dramatically different form. This broad-crowned tree grew free-standing at the edge of a pasture. It had, as a young tree, no neighbors growing close by. As is typical for a solitary tree, the crown gradually spread out broadly in all directions, attaining a relatively spherical shape. In general, branches grow outward and ramify into the space of greater brightness surrounding them. The leaves and branches themselves create darkness so that the outward spreading is toward greater brightness. A tree is, in part, its own context (see also Holdrege 2005a).

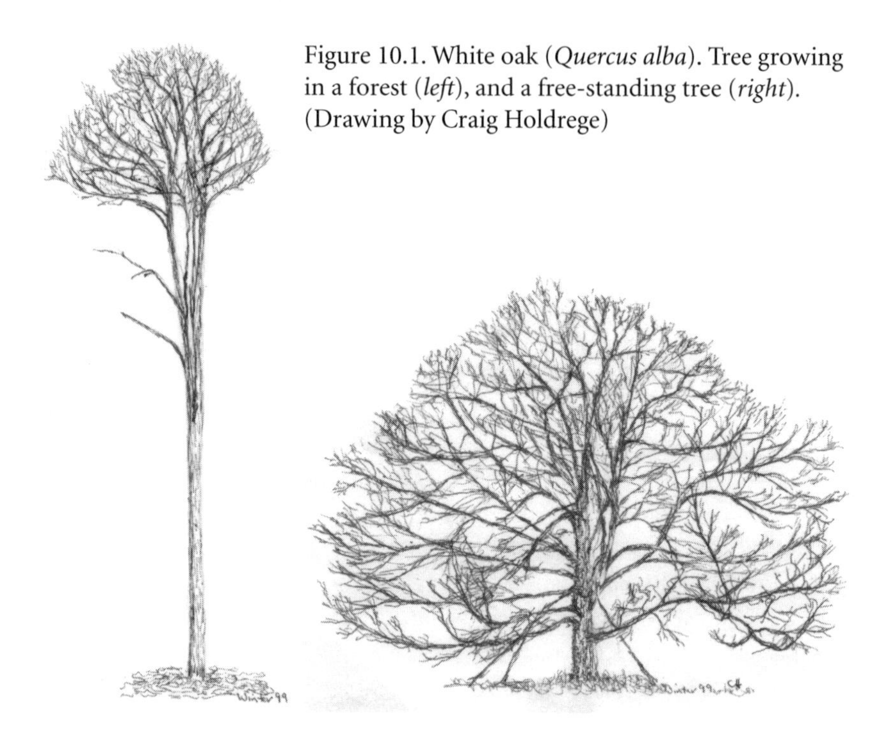

Figure 10.1. White oak (*Quercus alba*). Tree growing in a forest (*left*), and a free-standing tree (*right*). (Drawing by Craig Holdrege)

Context

Organisms such as these two white oaks teach us that we begin to understand isolated facts only when we look at them in the light of a larger context. Indeed, facts appear isolated only because we have abstracted them from all their relationships and connections in order to focus more clearly and narrowly. In order to gain real understanding, however, we need to overcome this isolation, re-creating the context in which life gains its fullness.

Although it may sound simple to restore context in order to gain understanding, it is not. Our contrary habits run deep. In science we are trained to seek the clear outlines of decisive facts held in sharp focus. We learn to look for underlying mechanisms—material causes—by excluding from view the ambiguities of any larger view. This approach is called reductionism, and it is taught, if not always by name, in schools and universities around the world.

For example, in pursuing the physiological mechanisms that cause a tree's trunk to grow long and narrow, we might study cell growth. This would lead us to cell metabolism. Finally, we may discover genes associated with cell proliferation or elongation. Such a study is perfectly justified and leads to detailed knowledge.

But problems arise when we forget the limitations imposed upon our conclusions by the narrowness of our focus. Our answers are valid only within the boundaries of our chosen methods and perspectives. We have not, after all, explained the tree's form when we determine some of the physiological parameters of cell growth. If we believe we have, then we have lost sight of the white oak growing in a bottomland forest community. Outside this context the physiological processes associated with elongation simply do not occur. Unfortunately, it is unlikely today that a scientist studying genes will know very much about the way trees grow. The requirements of specialization leave little time for cultivating the contextually rich knowledge of an organism in its different natural settings.

The more we analyze, the more detailed, but also the more fragmented, our knowledge becomes. We run the danger of treating the organs, tissues, cells, genes, or other substances we investigate as if they were entities unto themselves. If we do this, then we may begin considering the organism to be an agglomeration of independent parts. It is only a small step to the image of the organism as a mechanism, driven by the workings of its mechanically interacting parts.

This way of viewing organisms—and biological phenomena more generally—we call object-thinking. We object-think when we focus on a detail or part of a larger system and then proceed to treat this part as an independent entity, even when we are trying to reintegrate it into a larger whole. The consequence is a mechanistic view of life and organisms.

It is an altogether different matter when our analysis is guided by the intention to understand any given biological phenomenon in light of its larger context, as in the case of the oaks. In this approach parts do not obtain independent status, because from the outset they are viewed as members of a larger whole. We call this contextual thinking, or more generally, a contextual attitude.

Depending upon which approach we choose to take, we will arrive at very different pictures of an organism. And, moreover, the way we think about organisms affects the way we interact with them, as will become evident in considering the cow.

The Cow as Organism

> Hence we conceive of the individual animal as a small world, existing for its own sake, by its own means. Every creature has its own reason to be. All its parts have a direct effect on one another, a relationship to one another, thereby constantly renewing the circle of life. (Goethe 1995, 121, written in 1795)

The German poet and scientist Johann Wolfgang von Goethe spent decades cultivating a contextual approach to organisms. He recognized not only the importance of looking at the way plants and animals interact with their environments, but also—as the above quotation indicates—that the organism itself is a context in which we must understand the particulars of its morphology and physiology (see chapters 12 and 14). The following description of the cow was written in this spirit and is based in part on a masterful portrayal by Ernst-Michael Kranich (1995, 19–28).

When we drink milk or eat yogurt or cheese, we are consuming products that connect us to the work of countless people: the farmers and those who produce the actual products; those who market, sell, and distribute the products; those who devise and build machines for these activities; and so on.

All this activity radiates out from the cow—the primary source of our milk products. The cow in turn depends upon the meadow grasses and wildflowers to produce its milk. At the same time, the cow's dung fertilizes the plants from which it lives.

Cows are grazers. They live in the midst of the food they eat. The cow lowers its head to the ground and touches the meadow plants (or the hay in its stall) with the front end of its soft, moist snout. The cow does not bite off the plants with its teeth or lips, but reaches out with its rough, muscular tongue, enwraps the plants, and tears them off. It clearly needs to use its tongue in this way—cattle that receive soft feed begin to lick their fellow cows much more than usual. The tongue needs the stimulation of roughage.

After it has torn off a few portions and chewed a bit, the cow swallows a mouthful. This activity continues for a few hours. The food reaches the rumen, the huge first chamber of the four-chambered stomach. Occupying the entire left side of the abdominal cavity, the rumen can hold up to forty-five gallons.

Digestion in the rumen is facilitated by microorganisms that break down cellulose, the main, hard-to-digest component of roughage. Secretion of saliva, bacterial activity, and the muscle activity of the rumen are all stimulated by roughage. In fact, the rumen only finishes its development and becomes functional when a calf begins to feed on grass or hay.

When the rumen is about half-full, portions of the partially digested food are regurgitated back into the mouth. Rumination begins. Cows usually lie on the ground while ruminating. They grind their food between their large cheek teeth in rhythmical, circling motions of the lower jaw. You are probably familiar with the picture of calm presented by a herd of cows, lying in a meadow, their activity focused inwardly on grinding and digestion.

Digestion involves an intensive production, circulation, and secretion of body fluids. The process begins in the head. While the cow is ruminating, its salivary glands secrete copious amounts of saliva—up to forty gallons a day. The drier the feed (for example, hay), the more the saliva, and the greater the amount of water a cow drinks. As Kranich points out, functionally one can consider the mouth to be a fifth chamber of the stomach.

After rumination, the food is swallowed, entering first the other three chambers of the stomach and then the small intestine. In these organs, fluids are removed from the food and new digestive juices are secreted until finally the cow has broken down its food as far as it can.

Characteristic for cows is their fluid dung, in contrast to the solid dung of other ruminants, such as sheep or goats. The cow's large intestine does not absorb as much fluid out of this final section of the digestive tract. In fact, from its moist snout, through the whole digestive tract, and finally in its dung, the cow shows more fluidity than other ruminants.

The digestive process is related to the blood—a fluid organ that connects all organs of the body. For every quart of saliva, three hundred quarts of blood pass through the salivary glands. The other digestive organs are sustained by a similarly strong circulation.

The intensive transformation of substances and secretion of fluids characterizing the digestive process are heightened in the formation and secretion of milk. Substances produced by digestion are withdrawn from the blood in the udder. For every quart of milk, three to five hundred quarts of blood pass through the udder. Glands in the udder then create a wholly new substance—milk. This is not a substance that is used by the cow or excreted; rather, it serves another growing organism—the calf. The cow only begins to produce milk after she has given birth to a calf, and the calf has begun to suck on the teats.

When we build up a picture of the cow in this way, we begin to see the cow as a total organism. We view each part in the context of other parts, so that the animal as a whole comes into view, even if only in an elementary way. One result of this endeavor is that milk loses its isolated status as a product we consume. As consumers we tend to take for granted our relation to the cow. When we gain some insight into the cow, we can begin to see milk as the special substance it is, created by the interaction of the internal ecology of the cow's physiology with its food and environment.

The Cow as Bioreactor

Until this century the cow gave about as much milk per day as her calf would have drunk, had it not been weaned—about two to three gallons in present-day breeds (in India, about one-half gallon per day). In our time, the dairy cow's milk production can exceed seven gallons per day. This increase has taken place essentially within the last fifty years.

How has the increase been made possible? First, by breeding larger cows that by virtue of their size eat more, digest more, and give more milk. Second, by feeding them differently. When cows receive more high-protein grains in their feed, they produce more milk. But since, as we have seen, cows need roughage, this dietary change has its limits.

A simple method has been developed to circumvent the need for roughage in steers bred for beef (Loerch 1991). The steers are "fed" plastic pot scrubbers—the ones we buy in supermarkets—instead of roughage. In trials, pot scrubbers were wrapped in masking tape and then, one after another, eight scrubbers were pushed down the steer's throat into the rumen. The tape soon detached from the scrubbers, which "were observed to float on the surface of the ruminal contents in these steers and to form a mat similar to that observed when ruminants are fed roughage." The scrubbers remain in the rumen for life.

The trials indicated that steers fed 100 percent concentrate plus pot scrubbers grew at approximately the rate of cattle fed 85 percent concentrate with 15 percent roughage (corn silage). Evidently, the scrubbers stimulate the rumen walls in a manner similar to roughage. In undertaking his research, Loerch surmised that "because roughage is relatively low in energy and is expensive, it would be beneficial if roughage could be eliminated from cattle diets without sacrificing performance." It is by no means clear that a farmer would actually save money using this method, since it is not a given that 15 percent more concentrate would be cheaper than producing or buying a corresponding amount of corn silage. But some farmers or feedlots have evidently used Loerch's method, since, as a university animal scientist, he is reported to have received many phone calls "from bewildered butchers who have found pot scrubbers in the guts of slaughtered cattle" (*New York Times,* August 29, 1992).

In its starkness this example is illustrative. It shows not only how strongly the desire to lower costs is a determining factor in agricultural research, but also in what narrow terms the cow is viewed. The cow's need for roughage is reduced to a mechanical function, and this can be substituted for. The sensory qualities of hay or silage—taste, smell, texture—are not considered. Nutritional considerations are reduced to ascertaining that roughage is low-calorie feed and therefore not effective for fast growth. The steer can no longer ruminate because the scrubbers are too large to be regurgitated. Has this no significance for the animal's well-being and physiology? The cow as a mechanism and not the cow as an organism stands behind this roughage substitute.

Perhaps a more enlightened age will discover that the nutritional quality of foodstuffs such as milk or beef are dependent not only on the results of biochemical analysis, but also on the way the animals are raised and cared for. Coupling the view of the cow as a mechanism with a one-sided economic perspective that emphasizes cost-effectiveness has become increasingly prevalent in our times. This is particularly true in genetic engineering: "Producing human pharmaceutical proteins in the milk of transgenic livestock has been an attractive possibility. . . . Such 'molecular pharming' [*ph*-armacy + f-*arming*] technologies are appealing for a number of reasons. They offer the potential of extremely high volumetric productivity, low operating costs, and unlimited multiplication of the bioreactor [that is, the animal]. . . . In this issue of *Bio/Technology* three groups report significant progress in realizing these benefits. . . . Their results provide convincing demonstration of the feasibility of using animals as commercial bioreactors" (Bialy 1991).

The attempt to continually increase milk production reflects the treatment of cows as commercial bioreactors. This tuning of the bioreactor in a specific direction has brought with it some unwanted side effects. These include fertility problems, mastitis, and leg and hoof afflictions. For these and other reasons (Hadley et al. 2006), high milk-producing cows are often slaughtered after only three years of lactation (when they are five-year-old animals). Without the demand to produce as much milk as possible in a short period of time, a cow will reach its peak of milk production after three or four years of lactation, and will continue healthy lactation for a number of years beyond that. When we begin to think in terms of the organism, we learn to expect that the desired effect of our manipulations will in all likelihood be only one among many changes. From the point of view of the organism, there is no such thing as a side effect. The organism is a whole. If we change a part, the whole is changed, and this change will likely manifest in ways that go beyond any desired effects.

For example, mastitis can accompany increased milk production. Mastitis is an inflammation of the udder. Since it is an infectious disease, strict hygienic procedures are called for to prevent bacteria from entering the udder via the openings in the teats. But this is only one side of the problem. The symptoms of a classic inflamation—warmth, redness, pain, and swelling—are associated with an increase in the amount of blood flowing through the inflamed organ. During lactation, circulation through the udder is increased, and when milk production is pushed to the utmost degree, the udder is almost on the verge of inflammation without bacteria. The cow's physiology is stressed, and when bacteria do enter the udder, mastitis is likely.

In 1993, the Food and Drug Administration (FDA) approved the commercial sale of milk, milk products, and meat from cows treated with recombinant bovine growth hormone (rBGH). This hormone is produced by bacteria that have been genetically altered by a cow-derived DNA strand that is related to the animal's production of growth hormone. In some unknown way, growth hormone stimulates milk production. Cows injected with this hormone produce 10 to 20 percent more milk.

Much controversy has surrounded the use of rBGH, and in Europe its use has not been approved. The FDA was concerned solely with the product's safety. FDA scientists concluded that experimental evidence (provided by manufacturers of rBGH) demonstrates that milk from treated cows is in essence chemically identical to milk from untreated cows. Therefore, the FDA sees no reason for the milk to be labeled as coming from rBGH-treated cows.

Extensive testing of rBGH was done on rats as part of the FDA's procedure for establishing the safety of the substance. Although such experimental results cannot simply be assumed to be valid for cows, they are in and of themselves interesting. Researchers found that the whole organism was affected by rBGH. The treated animals were larger than normal. When the researchers investigated the individual organs, they found that some were proportionately smaller while others were proportionately larger than normal. Such changes depended in part on the animal's sex. "Ratios of organ weight to body weight were increased for spleen and adrenal [gland] and decreased for testes in male rats, and increased for heart and spleen and decreased for brain in the female rats" (Juskevich and Guyer 1990).

Such detailed analyses have not been performed on cows, but the question of the effects of rBGH has been a source of major controversy and concern. Monsanto, the sole producer of rBGH (which is sold under the brand name Posilac), claims that there are no significant side effects. Monsanto even markets Posilac as "one of the leading dairy animal health products in the United States" and maintains that it does not negatively affect cow health (http://www.monsantodairy.com/about/index.html). Some independent scientists have come to different conclusions. One research group analyzed Monsanto data and concluded that milk from rBGH-treated cows contained an average of 19 percent more white blood cells than milk from untreated cows (Millstone et al. 1994). White blood cells enter an organ as part of the inflammatory reaction. An increase in white blood cells is "associated with increased risk of mastitis."

In 2003, Canadian scientists published the results of an analysis of all peer-reviewed studies related to rBGH and also the data Monsanto submitted to the Canadian government (Dohoo et al. 2003a and 2003b). This comprehensive study, which was funded by the Canadian government, found that cows treated with rBGH not only produced more milk (11 percent to 15.6 percent more), but also ate on average 1.5 kilograms more dry matter (concentrate, hay, or silage) per day. This increase in food did not, however, compensate for the extra physiological exertion associated with higher milk production, so the cows tended to lose weight, and their overall body condition was not as optimal as in cows that received no rBGH. Their susceptibility to mastitis was greater by 25 percent, the risk of lameness increased by 55 percent, and the likelihood of failed conception rose by 40 percent. Clearly, treating the cow as a commercial bioreactor has its limits—the cow as an organism can only be pushed so far.

Out of what context is rBGH produced, and into what context do its effects radiate? More milk is produced in the United States than is consumed within the country or exported. Every year the U.S. Department of Agriculture purchases excess dairy products (mainly nonfat dry milk) from farmers as part of its Milk Price Support Program. In 2002, for example, nearly 7 million pounds of surplus dairy products were purchased by the government at a price of $617,000,000, not including storage costs (USDA 2004).

Given the overproduction of milk, which goes back decades, it is clearly absurd that a product that increases milk production has been aggressively marketed in the United States. Such a total separation of production from actual needs is a consequence of our sick economic system, which emphasizes growth, higher production, and cost-cutting above all other factors. In agriculture, this ideology has led to the development of ever larger factory farms. Higher productivity is achieved to the detriment of the connection (both physical and emotional) between farmers and the plants and animals upon which their work is based.

When a large chemical company invests many millions of dollars to develop a new product such as rBGH, it must aggressively market the product. Farmers (particularly those with large farms) who seek further mechanization and ever higher production are most likely to use rBGH. But others follow in fear of not being able to compete. An irrational, needs-divorced spiral of growth is the consequence.

And what about the cow? As long as we treat it as a commercial bioreactor, there is every reason to manipulate it in order to further increase milk production. Growth hormone-treated cows to support economic growth—what could be more logical? But the cow does not simply fit into this system. It is not a commercial bioreactor; it is an organism that has its own needs and its own boundaries. The increased incidence of disease and other maladies that accompany factory-farming and the use of rBGH are clear indicators. But industrial farming has lost sight of the cow.

Only by gaining insight into the cow as a "small world, existing for its own sake," can we recognize its specific characteristics and needs, and begin to fit our actions into its context. Practicing this point of view is made extremely difficult by current economic realities. This problem, where it is recognized, has led (to mention one example) to the establishment of Community Supported Agriculture (Groh and McFadden 1997; Henderson 1999). Here farmers and consumers enter an economic association that frees the farmer to some degree from the compelling necessity to increase production and lower costs. The consumer community provides the farmer with an income. At the same time, farm production is related more directly to consciously affirmed consumer needs. Within this setting it becomes possible to handle animals like the living, sentient beings they are.

Taking Responsibility for Our Point of View

The concept of responsibility is eminently contextual. In every action we connect ourselves with the world. The world then carries our imprint. In this fundamental way we are responsible for everything we do, whether we are aware of it or not. Most of us will not feel responsible for a tree that falls over in a windstorm. If, however, we had noticed that the tree trunk was rotting before the storm, we might have a bad conscience afterward and sense that it was irresponsible not to have cut down the tree, because someone might have been hurt. In the first case, we hold the tree's falling to be an "act of nature"—external to us and beyond our influence. In the second case, we find ourselves bound to the tree by our choices and their effects.

When we sense responsibility, we feel a connection between our thoughts, feelings, or actions and something in the world. We internalize what we might have left "out there." So long as we can hold things at a distance, no feeling of responsibility arises. By the same token, when we deny responsibility, we maintain or create an inner distance between ourselves and the other.

Our responsibility, then, is governed by those very qualities that constitute a living organism. Animals and plants do not relate to their surrounding environment as if it were an object. They internalize what was outside of them, integrating it into themselves. In so doing they change. Similarly, our feeling of responsibility emerges only where an "other" has to some degree ceased to be a mere object. We dwell within it, or it dwells within us.

This taking of responsibility begins with the awareness of our point of view. There is no compelling necessity to look at the world in a particular way. We choose our viewpoint and cannot deny that we are connected to the results of our choice (cf. Edelglass et al. 1997, 135–38).

Reductionism can be practiced responsibly only when we remain aware that, while it allows an exact focus and may lead to an understanding of particulars, it also destroys context. Recognizing this, we will not reify the abstract products of reduction into independent objects separate from us. Rather, we will know them to stand within a larger context of which we, and our point of view, are also a part.

As it is, reductionist science does not tend to cultivate such awareness. Symptomatic is a "straw poll" a biology professor made of first-year biology students "to see how many of them had come across the notion that, even in these banausic times, there is such a thing as the philosophy of science, and that reductionism constitutes one approach by which one may seek to understand nature. To this end, I asked them if they had come upon the term 'reductionism.' Not one of them had" (Kell 1991). It is essential to lead students into a conscious recognition of the specific approach they are pursuing. How can they take responsibility for their way of thinking, and then for the results of that way of thinking, if they know nothing of their own or other points of view? Through lack of critical reflection, reductionism perpetuates itself and ignores the roots of responsibility within the human being.

A contextual approach should not be looked at as yet another solidified doctrine or theory. Rather, it is a necessary complement to the prevailing conceptions and practices of contemporary science. It is a way to make science a healthier whole, modeled after the organisms it studies.

An awareness of this whole would lead to greater thoughtfulness and caution. We would ask larger questions about what we are doing, and place our actions into the context of their outward-rippling effects. We would try to accompany the effects of our actions out into the world. After all, no transgenic organism exists without us; we helped it into existence. Are we prepared to live side-by-side with it?

It is unrealistic to suppose that one day we will be able to predict before the fact all the consequences of any scientific or technological endeavor. Life always brings surprises, and we cannot foresee the future through the lens of previous knowledge. So it is all the more essential to imbue an undertaking with a contextual attitude from the outset. Involving the whole human being, this new way of looking will require wakefulness and the will to change.

Chapter 11

The Forbidden Question

Few of us would want to condemn a wolf to the life of a house pet. But we can be quite sure that sometime in the human past wolves, or similar wild creatures, were in fact domesticated. Was this a bad thing? It does not seem that a dog whose business is herding sheep or retrieving fowl has such a bad life. Nor do these dogs seem perverse additions to the planet's canine ranks.

On the other hand, what about the monkeys given jellyfish-derived genes by researchers at the Oregon National Primate Research Center, so that they will glow green under ultraviolet light? Or the genetically engineered salmon produced by Aqua Bounty Farms in Waltham, Massachusetts—salmon designed to live within the artificial confines of fish farms, where they grow much faster than normal salmon, and achieve a greater size, while consuming less food? Or what about the "pharmed" goats created by the Genzyme Corporation in Massachusetts and Nexia Biotechnologies in Canada—goats whose genes have been engineered to make the animal a kind of factory for the production of pharmaceutical chemicals or other useful substances harvestable from its milk? Does such a goat have a bad life? To judge from immediate appearances, it continues doing all the things goats normally do, from grazing to butting to bearing offspring. Do we have any grounds for questioning the work of the genetic engineers who contrived what they view as a "living factory"?

The answer depends not only on our own engineering goals, but also on the organisms being engineered. Who, or what, are we dealing with? Is there anything in the character, or nature, of the monkey, salmon, or goat that suggests our interventions in its destiny are wrong or inadvisable?

Here, however, we run into two entrenched prejudices. One is the biologist's longstanding distaste for any talk of an organism's "nature." The fear is that such language smuggles into science the spooky notion of a ghost in the machine, an immaterial, mystical "essence" directing the organism from within. The language smacks of a being—a who—and we intuitively take beings to be very different from machines. Many biologists are far more comfortable with the view approvingly summarized by philosopher Daniel Dennett: all life is founded upon "an impersonal, unreflective, robotic, mindless little scrap of molecular machinery"—machinery in which, as he puts it emphatically, "there's nobody home" (1995, 202–3).

The other prejudice that hinders our capacity to value nonhuman organisms on their own terms is expressed in the more recent conviction that whatever nature the organism actually does possess is defined arbitrarily and mechanistically by its genes. When we manipulate these genes, there is no external measuring rod, no standard, by which to assess whether our actions do violence to the organism's nature. We are simply exercising our godlike prerogative to impose our own preferences upon the organism—that is all. Who can judge whether one set of mechanisms is more justified than another, except by weighing how well the overall machine serves the particular purposes for which we are inclined to employ it?

"Genes are mechanisms of destiny, and questions of destiny are more and more given into our own hands"—there's at least a shadow of truth in this. But if we are to carry our burden of destiny responsibly, it cannot be right for us to remain naive about genes and the organisms that bear them. Above all, it cannot be right for us, when facing the wolf or goat or any other organism, to continue avoiding the question, "Who, if anyone—what sort of being—is there?"

Plastic Traits

In the early spring of 1939 a male goat was born in Holland without forelegs. During the first seven months of its life, while its skeleton and musculature were developing, it passed its days in a grassy field, moving forward by jumping on its hind legs in a semi-upright posture. According to the eminent Dutch morphologist E. J. Slijper (1942; 1946), who studied the bipedal goat, it moved much like a kangaroo, with both legs leaving the ground at the same time.

While this particular animal happened to die in an accident when it was one year old, such malformed mammals are not uncommon and are capable of reaching old age. Slijper, for example, mentions a two-legged dog that lived twelve years. Moreover, the evidence from these animals tells a consistent story: in adapting to its malformation, a young, developing animal can depart in astonishing and systematic ways from the "normal" structure of its species. In the case of the bipedal goat, it not only *behaved* in some ways rather like a kangaroo; it also *grew* rather like a kangaroo, with the changes penetrating right down into the form and articulation of its muscles and bones, as illustrated in figure 11.1.

When, even in the hard substance of its bones, a goat can begin to resemble a kangaroo, what has become of its fixed "goat nature"? We might like to think that *this* gene (or complex of genes) governs *that* trait. But the bipedal goat points us to a vast domain of evidence sug-

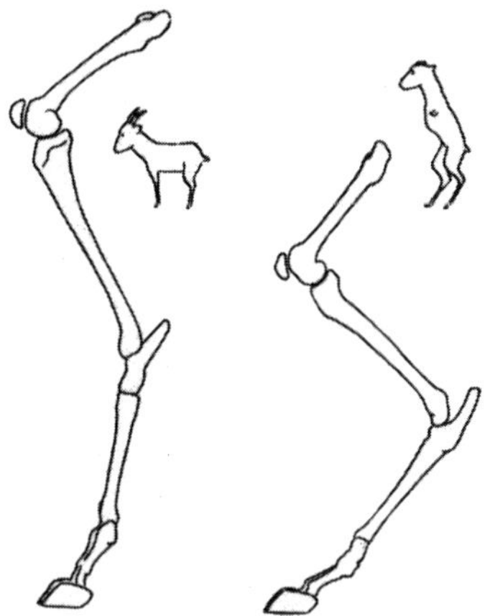

Figure 11.1a *(left)*. Leg bones of normal (left) and bipedal (right) goat. (Reprinted from Slijper 1946)

Figure 11.1b *(below)*. Spinal columns of normal (bottom) and bipedal (top) goat. (Reprinted from Slijper 1946)

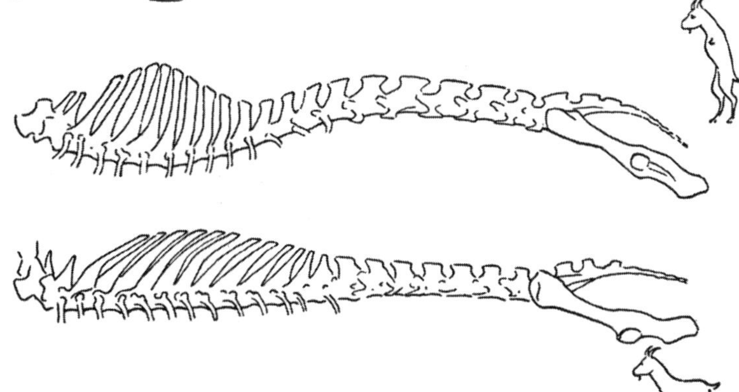

gesting *there is no neatly given "that"* to be determined. Traits are not rigidly fixed, definable things.

This conclusion does not rest upon freaks of nature. The bipedal goat's powers of adaptation are the same powers that produce, under more normal circumstances, the familiar goat morphology. What we take to be normal always reflects a common set of circumstances as much as anything else. Moreover, even unexceptional circumstances may yield forms that are far from being fixed and predictable. Figure 11.2, for example, shows some of the observed variations in the form of the human stomach and the human liver.

In her textbook *Developmental Plasticity and Evolution,* Mary Jane West-Eberhard remarks on the extreme variability even in the most vital organs. "Variants that might be considered monstrous pathologies if seen in isolation occur as part of normal development in human populations." This illustrates what she calls "phenotypic accommodation"—the "adaptive mutual adjustment among variable parts during development" (2003, 51). If, for whatever reason, one part changes drastically, the other parts accommodate themselves to the new conditions.

Even when the change is inflicted through violence from outside, the

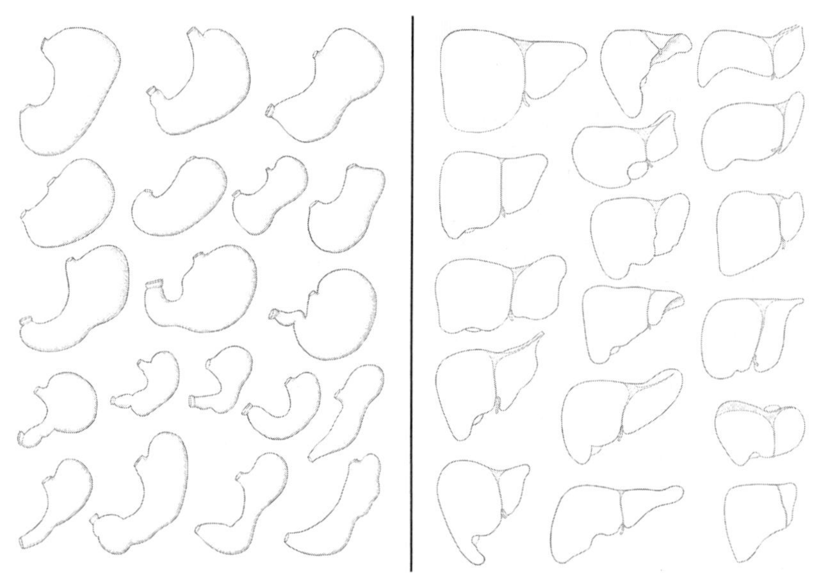

Figure 11.2. Variations in the size and shape of the human stomach (*left*) and liver (*right*). (Reprinted from Anson 1950, 287, 288)

organism may adapt with startling grace. West-Eberhard cites a classic, if grotesque, experiment—among many similar ones—where the eye of a large salamander is transplanted into the embryonic skull of a smaller species. The eye grows to its normal large size, and the host salamander "develops a proportionally larger cartilaginous optic cup to perfectly accommodate the larger eye, as well as a concordant change in the tectal neuron population of the midbrain corresponding to an increase in the number of retinal ganglion cells associated with the grafted eye" (54).

The leg bones of goats and optic cups of salamanders do not look like good candidates for well-defined traits. Traits are plastic, and what we actually see always reflects the organism's impressive powers of response to its circumstances.

Is Anyone There?

The formulaic notion that "this gene governs that trait" suffers from more than our inability to specify *that trait.* As we have seen in previous chapters, we are also unable to nail down *this gene* as a well-defined and self-contained entity. This is hardly surprising, since the gene was originally conceived by Gregor Mendel as an unknown "factor" directly correlated with a given trait. Where the well-defined trait cannot be found, it stands to reason that no fixed, single factor corresponding to the trait can be found either. You can't explain plastic traits by appealing to rigid determining factors.

But if there are no well-defined traits to be determined and no well-defined genes to do the determining, are we not left with a kind of chaos where anything can happen? It begins to look as though one of the entrenched prejudices mentioned earlier is well justified: it doesn't make much sense to speak of an organism's *given nature* when everything seems to be in such flux.

But perhaps we have been misled by our insistence upon seeking mechanisms of control where none is to be found. Another way of looking for unity and coherence would lead us to put our question in the terms of a different tradition—not "what is the controlling mechanism?" but rather "What sort of being is there?"

Of course, this brings us right up against prejudice again. "Is anyone there?" has become a taboo question for today's scientist. It's true that people who actually befriend and get to know animals—or, at least, higher animals—seem inevitably to violate the taboo by addressing the animal as if someone were there. But the cool-headed molecular biologist in her antiseptic, stainless-steel laboratory is hardly tempted to flirt with such a notion as she manipulates excised tissues and chemical isolates. Nevertheless, if the organism, faced with constraints not only from the environment but also from its own body, responds coherently to these constraints in its own distinctive and characteristic manner, nothing requires us to ignore the scientific questions raised by this fact. Nothing prevents our asking, as scientists, *who* or *what sort of being* is doing the responding? What consistency of character can we actually observe?

If traits are plastic and genes are not simple, predictable mechanisms, and if what we actually see reflects the organism's impressive powers of response, then doesn't it make sense, when we want to understand a particular organism's nature, to try to understand these powers themselves rather than some fixed set of determinants? Observation of this or that feature of an organism is important, but its significance lies in the fact that it has crystallized, under one set of circumstances, from a range of possibilities governed by an adaptive power. It speaks of a living origin that, like all life, exhibits both integral coherence and adaptive plasticity.

The fact that traits are not definitively given does not mean the organism lacks recognizable character. It's just that the recognizable character we find is qualitative, expressive, and dynamic. Goats can, under extreme circumstances, develop certain kangaroo-like features. But goats no more become kangaroos than an eagle forced to live on the ground becomes a chicken. Even in their most radical adaptations, both goat and eagle give recognizable expression to their own nature.

To recognize that nature, however, requires us to develop the appropriate, artistically disciplined power of seeing. The search for mechanisms only blinds us for *this* seeing. We must loosen our perception from the immediately given physical object—the particular form of leg bone or spine, locomotion, or grooming—and discover what lives and moves as a unique, qualitative, and adaptive shaping power working through the entire organism, adjusting part to part.

"Entire organism" and "way of being": there is no need to make more of these terms than is justified by our current powers of observation. Let them remain as open as necessary. But neither should we prejudice science by deriding or ignoring the observations from which the terminology naturally arises. This means, among other things, not prejudging what sort of observation might yield fruitful results. It might be that asking about an organism's governing unity, its way of being, requires us to exercise our faculties in an imaginative—a qualitative, pictorial, and dynamic—manner. It might be that asking about the nature of a creature is more like asking about the organic unity of a work of art than asking about the mechanical and external relations of the artwork's parts.

Seeing with Moral Earnestness

We can and do, as human beings, choose to modify plants and animals for human purposes. If this interaction is to be at all responsible, we cannot do this solely according to our own sense of utility. At least to some degree, we must get to know the organism we are dealing with on its own terms—that is, by attending to how it expresses its unique qualities through its form, life, and behavior. Only then can we adapt our intentions to its propensities. Otherwise, we merely manhandle it.

Moral judgments are determined in part by the *nature* of what is there, which is exactly as it should be. We cannot find a moral relation to anything except insofar as we encounter a distinctive way of being with its own character. Our sense of obligation will be radically shaped by this character. It is hardly the same thing to deal with a monkey or an ant, a redwood or a blade of grass. The point here isn't to elevate the monkey and denigrate the ant, but only to say that moral response must adapt itself to the distinctive character of what is there. It's very different to cut huge numbers of blades of grass than to cut a swath of redwoods, if only because of the regenerative capabilities of the two species.

We do not need to choose between arbitrary manipulation on the one hand and the pretense that we can live without affecting the destiny of our fellow creatures on the other. No living organism can exist in perfect isolation. Between the detachment of cold manipulation and that of disconnection lies another option: responsible engagement. That is, we can enter into mutually respectful conversation with the other inhabitants of the earth. Just as each of us unavoidably influences the people around us and is shaped by them, so it is with all creatures on the planet.

The idea that nature presents us with partners in conversation meets with strong resistance in many corners of society—not just scientific laboratories. Our intensifying history of scientifically supported manipulation of nature, from wholesale habitat destruction to factory farms to the arbitrary shuffling of genes between species, is proof enough. Until we can lift the taboo against asking "Who is there?" our fellow organisms will remain strangers to us.

How alien they now are is evident when we look at some of the ways in which, for instance, genetic engineers have altered pigs: pigs that grow faster and have more lean pork, but suffer from lameness, gastric ulcers, dermatitis, and other maladies; pigs that glow green with jellyfish genes as a tool to detect gene expression; pigs richer in omega-3 fatty acids to make their pork more healthy for human consumption; pigs that excrete less phosphorus in their feces, so that we can continue to raise them in factory farms but reduce the pollution caused by dumping their manure into huge lagoons that seep into groundwater; and pigs that go blind due to retinal degeneration as models for studying human retina degeneration. Whatever you think about the vices or virtues of a particular modification, one thing is clear: in these experiments no one was thinking about the pig and its needs as a being to be reckoned with.

Given the apparent disinterest of scientists in what animals can tell us, it is no surprise that they can say little about the wisdom or folly of a biotech pig, a pharmed goat, or a re-engineered salmon. The scientific mind has spent a long time learning how to avoid looking in a way that might answer our questions.

In some ways, the avoidance seems to be getting worse. The older breed of naturalists continues to die out as young biology students are drawn toward the glittering hope of the genome-sequencing laboratories. And now, with so few researchers capable of getting to know and understand any given organism, there is a serious effort to substitute DNA "barcoding" for the actual characterization of newly discovered specimens. One can imagine a time when a machine will be able to note certain differences between two tissue samples with wonderful precision, and the question, "Who is there?" will seem little more than the memory of an ancient dream.

However, as long as a true scientific spirit still lives in us there is hope that at least a few adventuresome researchers will extend their observational reach so as to embrace long-avoided realities of the world. This has always been the way of scientific advance. Surely we can overcome the metaphysical dogma that says, "The nature of the organism has already been revealed to us as that of mechanism," and surely we can substitute for this faith a spirit of openness to whatever may reveal itself through the apparent unities of being we discover all around us. In the end, we may hope, a healthy interest will win out over our fear of the unexpected characters we may meet in this diverse and ever-unexpected world.

Meanwhile, genetic modification and various other technologies for intervening where nature and fate once ruled are now increasingly being applied to human beings. The unborn whose genes are scanned; the elderly kept alive, at least physically, by ever more sophisticated machines; the seriously ill who receive experimental genetic treatments; the embryos rich in desirable stem cells—on every hand we are being given the opportunity to intervene in processes once reserved for "human destiny," just as we have long and unthinkingly altered the destiny of so many of our fellow creatures. And for ourselves, just as for earth's other inhabitants, our failure to ask, "Who is there?" is already an answer—an answer that may prove a self-fulfilling prophecy: "No one."

Chapter 12

What Does It Mean to Be a Sloth?

> One more defect and they could not have existed.
>
> —George Louis Leclerc, Comte de Buffon

> Hence we conceive of the individual animal as a small world, existing for its own sake, by its own means. Every creature is its own reason to be. All its parts have a direct effect on one another, a relationship to one another, thereby constantly renewing the circle of life; thus we are justified in considering every animal physiologically perfect.
>
> —Johann Wolfgang von Goethe

We are losing animals. Not only numerically through the extinction of species, but we also are losing them in our understanding. Perhaps it might be better to say we've rarely taken animals as whole, integrated beings seriously, and therefore they have never really come into view for us. For that reason our scientific and technological culture can so casually manipulate what it does not know. The moment we get to know something more intimately, the less likely we are to treat it in a purely utilitarian fashion.

Imagine a biotechnologist wondering what causes the sloth to be slow and pondering whether the animal could be mined for "slothful" genes that might be put to therapeutic use in hyperactive children. Or another who wonders whether the sloth might be a good research model for testing the efficacy of genes from other organisms that enhance metabolic activity. As far as we know, no such research projects are in progress or being planned. But how easy it is to come up with ideas that hover in splendid isolation above any deeper concern with the animal itself! As human beings, we are intrigued and motivated by the seemingly boundless limits of doing the doable and do not feel limited by ignorance of what we're dealing with.

This chapter is an attempt to show how we can take steps to overcome some of that ignorance—of which we should nonetheless always be mindful—by beginning to grasp something of the organic lawfulness inherent in one animal, the sloth. With all its unique and unusual features, the sloth almost seemed to be prodding us to understand it in an integrated, holistic way. The poet and scientist Johann Wolfgang von Goethe, whose approach is described further in chapter 14, set the stage for a sound holistic approach to studying animals, and others have developed his method further.* Their work has influenced and inspired this study.

The Sloth in Its World

Even if you were to look hard and make lots of noise, you would most likely not see the most prevalent tree-dwelling mammal in Central and South America's rain forests. The monkeys scurry off and perhaps scream. The sloth remains still and hidden.

The rain forest is a tropical ecosystem characterized by constancy of conditions. The length of day and night during the year varies little. On the equator there are twelve hours of daylight and twelve hours of night 365 days a year. The sun rises at 6 A.M. and sets at 6 P.M. Afternoon rains fall daily throughout most of the year. The air is humid (more than 90 percent) and warm. The temperature varies little in the course of the year, averaging 25 degrees Celsius (77 degrees Fahrenheit).

Except in the uppermost part of the canopy, it is dark in the rain forest. Little light penetrates to the floor. The uniformity of light, warmth, and moisture—in intensity and rhythm—mark the rain forest. And it is hard to imagine a rain forest dweller that embodies this quality of constancy more than the sloth. From meters below, the sloth is sometimes described as looking like a clump of decomposing leaves or a lichen-

* For general expositions of Goethe's method see: Bortoft 1996; Goethe 1995; Steiner 1988. For the biological application of a holistic methodology see: Portmann 1967; Schad 1977; Schad, ed., 1983; Riegner 1993 and 1998; Kranich 1995 and 1999; Suchantke 2001 and 2002; Holdrege 1998, 2004a, and 2005b.

covered bough. The sloth's hair is long and shaggy, yet strangely soft. The fur is brown to tan and quite variable in its mottled pattern. Especially during the wettest times of the year, the sloth is tinted green from the algae that thrive on its pelage, which soaks up water like a sponge (Aiello 1985).

Since the sloth moves very slowly and makes few noises, it blends into the crowns of the rain forest trees. It took researchers many years to discover that up to seven hundred sloths may inhabit one square kilometer of rain forest (Sunquist 1986). Only seventy howler monkeys inhabit the same area.

The sloth spends essentially its whole life in the trees. It hangs from branches by means of its long, sturdy claws, or sits nestled in the forks of tree limbs. The contrast to terrestrial mammals in respect to orientation is emphasized by its fur. Instead of having a part on the mid-back, with the hair running toward the belly, as is typical for terrestrial mammals, the sloth's fur has a part on the mid-belly and the hair runs toward the back.

The sloth moves slowly through the forest canopy—from a few feet to rarely a few hundred feet in twenty-four hours. On average, sloths are found to move during seven to ten hours of the twenty-four-hour day (Sunquist and Montgomery 1973). The remaining time sloths are asleep or inactive. (Resting is the term often used to describe the sloth's inactive periods, but this isn't a sloth-appropriate expression. From what activity is the sloth resting?)

Limbs and Muscles

The sloth's ability to hang from and cling to branches for hours on end is related to its whole anatomy and physiology. The sloth is about the size of a large domestic cat. It has very long limbs, especially the forelimbs (see figure 12.1). When hanging, the sloth's body appears to be almost an appendage to the limbs. Feet and toes are hidden in the fur. Only the long, curved, pointed claws emerge. The toe bones are not separately movable, being bound together by ligaments, so that the claws form one functional whole, best described as a hook.

The two different genera of sloths are named according to the number of claws they possess: the three-toed sloth (*Bradypus*) has three claws on each limb; the two-toed sloth (*Choloepus*) has two claws on the forelimb and three on the hind limb. (There are many differences in detail between these two groups of sloths. Most of the specific informa-

Figure 12.1. The three-toed sloth. (Sketch by Craig Holdrege)

tion referred to in this chapter pertains to the three-toed sloth, unless otherwise indicated.)

With its long limbs the sloth can embrace a thick branch or trunk, while the claws dig into the bark. But the sloth can also hang just by its claws on smaller branches, its body suspended in the air. A sloth can cling so tenaciously to a branch that researchers resort to sawing off the branch to bring the creature down from the trees.

All body movements, or the holding of a given posture, are made possible by muscles, which are rooted in the bones. Muscles work by means of contraction. While clinging, for example, some muscles in the limbs—the retractor muscles—are contracted (think of your biceps) while other muscles—the extensor muscles—are relaxed (think of your triceps). When a limb is extended (when the sloth reaches out to a branch) the extensor muscles contract, while the retractor muscles relax. All movement involves a rhythmical interplay between retractor and extensor muscles.

It is revealing that most of a sloth's skeletal musculature is made up of retractor muscles (Goffart 1971; Mendel 1985a). These are the muscles of the extremities that allow an animal to hold and cling to things and also to pull things toward it. Its extensor muscles are smaller and fewer in number. In fact, significant extensor muscles in other mammals are modified in the sloth and serve as retractor muscles. A sloth can thus hold its hanging body for long periods of time. It can even clasp a vertical trunk with only the hind limbs and lean over backward 90 degrees with freed forelimbs. As the sloth expert M. Goffart points out, "in humans this feat is exceptional enough to be shown in a circus" (Goffart 1971, 75).

At home as it is in the trees, the sloth is virtually helpless on the ground. Lacking necessary extensor muscles and stability in its joints, a sloth on the ground can hardly support its weight with its limbs. Researchers know little about natural terrestrial movement of sloths. But on rough surfaces captive sloths have been observed slowly crawling (Mendel 1985b). If they are placed on a smooth surface, such as concrete, their limbs splay to the side. In this position a sloth can only drag its body by finding a hold with the claws of its forelimbs and pulling itself forward, using its strong retractor muscles.

Since the sloth's main limb movements involve pulling and the animal's limbs do not carry the body weight, it is truly a four-armed and not a four-legged mammal. The hands and feet are essentially continuations of the long limb bones, ending in the elongated claws, and do not develop as independent, agile organs as they do, say, in monkeys. We can also understand the dominance of the retractor muscles from this point of view. The human being, in contrast to most mammals, has arms as well as weight-bearing legs. The arms are dominated by retractor muscles, while the legs have more extensor muscles. Moreover, the arm muscles that move the arm toward the body are stronger than the antagonistic arm muscles that move the arm away from the body. This comparison shows us that the tendency inherent in the arm—the limb that does not carry the body's weight—permeates the anatomy of the sloth.

A sloth becomes quite agile if the forces of gravity are reduced, as in water. In water a body loses as much weight as the weight of the volume of water it displaces (Archimedes' Law). The body becomes more buoyant, and in the case of the sloth, virtually weightless. "Remarkably, sloths are facile swimmers. . . . They manage to move across water with little apparent effort. Where the forest canopy is interrupted by a river or lake, sloths often paddle to new feeding grounds. With no heavy mass to weigh them down, they float on their buoyant, oversized stomachs" (Sunquist 1986, 9).

With its long forelimbs the sloth pulls its way through the water, not speedily, but in a "beautifully easy going" manner (Bullock, quoted in Goffart 1971, 94).

On the whole, sloths have little muscle tissue. Muscles make up 40 to 45 percent of the total body weight of most mammals, but only 25 to 30 percent in sloths (Goffart 1971, 25). One can understand how the reduction of weight in water allows them to be less encumbered in movement. Sloth muscles also react sluggishly, the fastest muscles contracting four to six times more slowly than comparable ones in a cat. In contrast, however, a sloth can keep its muscles contracted six times longer than a rabbit (Goffart 1971, 69). Such anatomical and physiological details reflect the sloth's whole way of being—steadfastly clinging in a given position, only gradually changing its state.

The tendency to reduced muscle tissue can also be found in the head. There are fewer and less complex facial muscles (Naples 1985). Through its facial markings the sloth has an expressive face, but this is the expression of a fixed image, rather than expression through movement, since the facial area itself is relatively immobile. The outer ears are tiny and are essentially stationary. The sloth alters the direction of its gaze by moving its head, not its eyeballs. This rather fixed countenance is dissolved at the lips and nostrils, which, as the primary gateways to perceiving and taking in food, are quite mobile.

Paced Metabolism and Fluctuating Body Temperature

Since sloths are externally inactive or asleep for a good portion of the twenty-four-hour day and spend the remaining time slowly moving and feeding, they perform about 10 percent of the physiological work of other mammals of similar size (Goffart 1971, 59). All metabolic processes are markedly measured in tempo. Sloths use little oxygen and breathe slowly, and the respiratory surface of their lungs is small.

All metabolic activity produces warmth. Warmth is also needed for activity, for example, in the exertion of muscles, which in turn results in more warmth production. Birds and virtually all mammals not only produce warmth, but also maintain a constant body temperature. This is a striking physiological feat. A warm-blooded (endothermic) animal is like a radiating, self-regulating center of warmth. Warmth constantly permeates the whole organism.

Most mammals maintain a constant core body temperature of

about 36 degrees Celsius (98 degrees Fahrenheit), which changes very little despite variations in environmental temperatures. For example, in a laboratory experiment a mouse's internal temperature changes only four-tenths of one degree Celsius when the outer temperature rises or falls twelve degrees (Bourlière 1964). Exceptionally, however, a sloth's body is not so permeated by warmth; in other words, it is not constantly prepared for activity. Its body temperature can vary markedly.

Gene Montgomery and Mel Sunquist, who did extensive field research in Panama on the ecology and physiology of sloths, found that the sloth's body temperature fluctuated with the ambient temperature (Montgomery and Sunquist 1978). During the morning, as the ambient temperature rises, the body temperature rises. When found on sunny days, sloths were often on an outer branch, belly-side up and limbs extended, basking in the sun. Body temperature usually peaks at about 36–38 degrees Celsius soon after midday. It then begins to fall, reaching a low point of about 30–32 degrees Celsius in the early morning. The body temperature is usually about seven to ten degrees Celsius higher than the ambient temperature.

Although sloths are often active at night, their body temperature does not rise in connection with their increased activity. This shows, in contrast to other mammals, that the sloth's body temperature is less affected by its own activity than by the ambient temperature. According to Brian McNab (1978), the sloth "almost appears to regulate its rate of metabolism by varying body temperature, whereas most endotherms [warm-blooded animals—mammals and birds] regulate body temperature by varying the rate of metabolism." Raising the outer temperature under experimental conditions is, as Goffart puts it, an efficient way of "'deslothing' the sloth," since it then moves around more actively. But if its temperature remains at 40 degrees Celsius for too long, it can prove fatal.

A three-toed sloth has an insulating coat of fur comparable to that of an arctic mammal, which seems at first rather absurd for a tropical animal. It has, like an arctic fox, an outer coat of longer, thick hair and an inner coat of short, fine, downy fur. These allow the sloth to retain the little warmth it creates through its metabolic processes. But, characteristically, the sloth cannot actively raise its body temperature by shivering as other mammals do. Shivering involves rapid muscle contractions that produce warmth.

Clearly, the sloth is at home in the womb of the rain forest, which keeps constant conditions like no other terrestrial ecosystem. Not only by virtue of its coloring and inconspicuous movements does the sloth blend into its environment, but through its slowly changing body temperature as well.

Feeding and Orientation

Moving unhurried through the crown of a tree, the sloth feeds on foliage. We usually think of leaf eating as an activity done on the ground by mammals, such as deer. There are, in fact, relatively few leaf-eating mammals in the crowns of trees, and the sloth epitomizes them. Leaves are an abundant and constant source of food, and plants need not be chased down. Sloths are literally embedded in and surrounded by their food at all times and in all directions. Tropical trees do lose their leaves, but not all at once. Sometimes a tree may lose leaves on one branch, while it sprouts new ones on others.

Sloths don't eat just any leaves. They seem to prefer younger leaves, and each individual animal has its own particular repertoire of about 40 tree species from which it feeds (Montgomery and Sunquist 1978). A young sloth feeds together with its mother, often licking leaf fragments from the mother's lips. After its mother departs the juvenile at the age of about six months, the young sloth continues to feed from those species it learned from its mother. This specificity is probably a major factor in the inability to keep three-toed sloths alive in zoos. They usually die of starvation after a short period of time. In contrast, the two-toed sloth is more flexible and survives well in captivity, eating assorted fruits and leaves.

A sloth does not search for leaves with its eyes. Its eyesight is very poor, and it is shortsighted (Goffart 1971, 103–7; Mendel et al. 1985). The eyes lack the tiny muscles that change the form of the lens to accommodate for changing distances of objects. As if to emphasize the unimportance of its eyes, the sloth can retract them into the eye sockets. The pupils are usually tiny, even at night. Clearly, a sloth does not actively penetrate its broader environment with its vision, as do most arboreal mammals, such as monkeys.

The sloth makes little use of sight and hearing, relying much more on a sense that entails drawing the environment into itself: the sense of smell.

> I placed a sloth, hungry and not too disturbed, on an open area under the bamboos, and planted four shoots twenty feet away

> in the four directions of the compass. One of these was *Cecropia* [a primary food of three-toed sloths] camouflaged with thin cheesecloth, so that the best of eyesight would never identify it, and placed to the south, so that any direct wind from the east would not bring the odor too easily. The sloth lifted itself and looked blinkingly around. The bamboos thirty feet above, silhouetted against the sky, caught its eye, and it pitifully stretched up an arm, as a child will reach for the moon. It then sniffed with outstretched head and neck, and painfully began its hooking progress toward the *Cecropia*. . . . Not only is each food leaf tested with the nostrils, but each branch. . . . (Beebe 1926, 23)

So we should not imagine a sloth looking at its food. Rather, a sloth immerses and orients itself in a sea of wafting scents.

When the sloth is in the immediate proximity of leaves it feeds on, it will hook the branch with the claws of a fore or hind limb and bring the leaves to its mouth. Having no front teeth (incisors), it tears off the leaves with its tough lips, then chews the leaves with its rear, peg-like teeth. Unlike most leaf-eating mammals, the sloth lacks many deeply rooted, hard, enamel-covered grinding teeth. The sloth also has comparatively few teeth (eighteen compared to thirty-two in most deer). The teeth wear easily, but in compensation grow slowly throughout the animal's life. There is no change of teeth from milk to permanent dentition. Growth and wear are in balance.

While feeding, the sloth is continuously chewing and simultaneously moving food backward with its large tongue in order to swallow. Sloths can feed in all positions, even hanging upside down. A young, captive two-toed sloth showed "decided preference for eating upside down in the manner of adult sloths at eight months" (Goffart 1971, 114).

The sloth can move its head in all directions, having an extremely flexible neck. Imagine a sloth hanging from all four legs on a horizontal branch. In this position the head looks upward (like when we lie in a hammock). Now the sloth can turn its head—without moving the body—180 degrees to the side and have its face oriented downward. As if this were not enough, the sloth can then move its head vertically and face forward—an upright head on an upside down body (figure 12.2)! When it sleeps, a sloth can rest its head on its chest.

The sloth's neck is not only unique in its flexibility, but also in its anatomy. Mammals have seven neck (cervical) vertebrae. The long-

Figure 12.2. The three-toed sloth. Note the orientation of the head. (Sketch by Craig Holdrege)

necked giraffe and the seemingly neckless dolphin—to mention the extremes—both have seven cervical vertebrae. This fixed mammalian pattern is abandoned by only the sloth and the manatee. The three-toed sloth usually has nine, and the two-toed sloth has between six and nine cervical vertebrae. (The manatee has six.)

Centered in Its Stomach

Digestion in the sloth occurs at an incredibly slow rate. In captive animals "after three or six days of fasting the stomach is found to be only slightly less full" (Britton 1941). Leaves are hard to digest and not very nutrient-rich, consisting primarily of cellulose and water. Only with the help of microorganisms in the stomach can the sloth digest cellulose, breaking it down into substances (fatty acids) that can be taken up by the bloodstream.

The sloth's stomach is four-chambered, like those of ruminants (cows, deer, and so on), and is clearly the center of the digestive process. The stomach is enormous relative to the animal's overall size. It takes up most of the space of the abdominal cavity and, including contents, makes up 20 to 30 percent of total body weight. Nonetheless, digestion takes a long time. On the basis of field experiments, Montgomery and Sunquist (1978) estimate that it takes food about ten times longer to pass through a sloth than through a cow. Moreover, the sloth also digests less of the plant material than most other herbivores.

Through its stomach a mammal senses hunger. Most grazing mammals spend a large part of their time eating, so that food is continuously passing through their digestive tract. The sloth is, once again, an atypical herbivore since it feeds for a comparatively small portion of its day. A small rain forest deer, the same size as a sloth, ate six times as much during the same period of time (Beebe 1926). The howler monkey, which also lives in the canopies of Central and South American rain forests and whose diet comprises only about 50 percent leaves, eats about seven times as many leaves as do sloths. With its slow metabolism and digestion, the sloth's stomach remains full, although the animal eats so little.

As a stark contrast, we can think of carnivores like wolves or lions that regularly, as a normal part of their lives, experience empty stomachs. Their hunting drives are directly connected to their hunger. Hunger brings about the maximum aggressive activity of these animals. When a lion has gorged itself on forty pounds of meat, it becomes lethargic and sleeps for an extended period of time. The sloth's constantly full stomach is a fitting image for its consistently slow-paced life as well as, it seems, a physiological condition for it: "starvation makes [sloths] hyperactive" (Goffart 1971, 113).

After about a week of feeding, sleeping, and external inactivity, a change occurs in the sloth's life. It begins to descend from its tree. Having reached the forest floor, it makes a hole in the leaf litter with its stubby little tail. It then urinates and defecates, covers the hole, and ascends back into the canopy, leaving its natural fertilizer behind. (The two-toed sloth has no tail and leaves its feces lying on the leaf litter.)

The feces, the product of sloth metabolism, decomposes very slowly. The hard pellets can be found only slightly decomposed six months after defecation. Normally, organic material decomposes rapidly in the warm and moist conditions of the rain forest. For example, leaves decompose within one or two months (a process that can take a few years in a temperate-climate forest). Ecologically, sloth feces "stands out as a long-term, stable source [of nutrients] . . . and may be related to stabilizing some components of the forest system. . . . Sloths slow the normally high recycling rates for certain trees" (Montgomery and Sunquist 1975, 94). Sloths contribute not only slow movement to the rain forest but slow decomposition as well.

It is estimated that a sloth can lose up to two pounds while defecating and urinating, more than one fourth of its total body weight (Goffart 1971, 124). If one imagines a sloth with a full stomach (which it always seems to have) just prior to excreting, then more than half of its body weight is made up of its food, waste, and digestive organs. This quantitative consideration points to the qualitative center of gravity in the animal's life. But the sloth's stomach is more like a vessel that needs to remain full than a place of intensive muscular activity, secretion, mixing, and breaking down, as it is in the cow, for example.

Stretching Time

The naturalist William Beebe wrote in 1926: "Sloths have no right to be living on this earth, but they would be fitting inhabitants of Mars, whose year is over six hundred days long." Beebe was deeply impressed by the way in which sloths "stretch" time, another way of characterizing their slowness. We have seen how this quality permeates every fiber of their day-to-day existence. It is therefore not so surprising to find that the development of sloths takes a long time.

Sloths have a gestation period of four to six months, compared to a little over two months in the similarly sized cat. One two-toed sloth in a zoo gave birth after eight and a half months. Initially more surprising was the rediscovery of a female sloth in the rain forest fifteen years after she had been tagged as an adult. This means she was at least seventeen years old, "an unusually long life span for such a small mammal" (Montgomery, quoted in Sunquist 1986, 10). Thus, regarding time, the qualities of the sloth certainly speak a unified language.

Gravity and the Skeleton

If we look for the embodiment of fixed form in the organ systems of a mammal, then we come to the skeleton. The bony skeleton gives the mammal its basic form and is the solid anchor for all movement. The limb bones develop their final form in relation to both gravity and their own usage. An injured quadruped mammal will lose bone substance in the leg it is not using, which does not carry any weight. Conversely, in the other three limbs bone matter is laid down to compensate for the increase in weight carried and muscular stress.

The sloth has a special relation to gravity. As mentioned earlier, the limbs hold the hanging body; they do not carry it (figure 12.3). The sloth gives itself over to gravity rather than resisting it and living actively within it via its skeletal system. A sloth kept on the ground in a box developed raw feet from the unaccustomed pressure (Beebe 1926).

The other pole in relation to gravity is represented by hoofed mammals, like deer, horses, and giraffes. By virtue of their skeletal architecture they can relax their muscles and even sleep while standing. Their legs are solid, stable columns that carry the body's weight (figure 12.4).

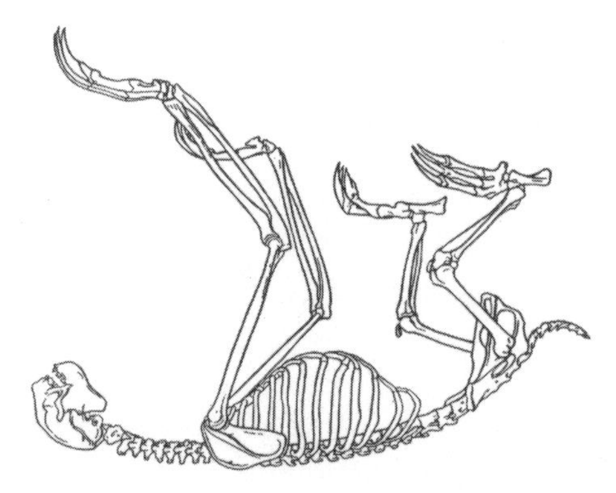

Figure 12.3. Skeleton of a three-toed sloth. (Reprinted from Young 1973, 600)

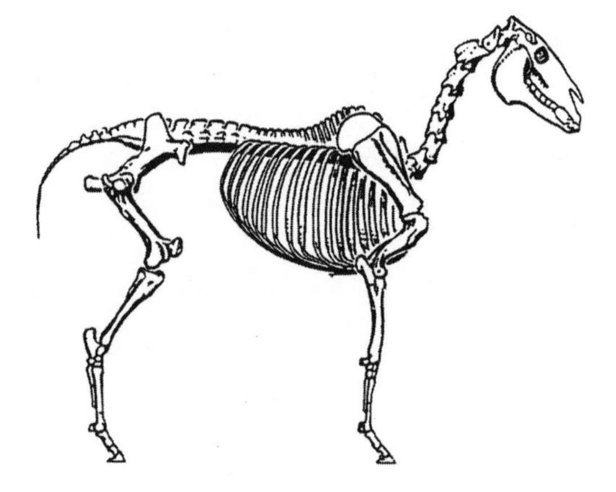

Figure 12.4. Skeleton of a horse. (Reprinted from Tank 1984, 108)

In contrast, the sloth has very loose limb joints. In his detailed study of the limbs of the two-toed sloth, Frank Mendel (1985a, 159) points out how unusual the "poorly reinforced and extremely lax joint capsules" are. This anatomical peculiarity allows a wide range of limb movement and is connected with the fact that the joints are not subject to compression as they are in weight-bearing limbs. Through clinging and hanging, the joints of a sloth are being continually stretched. Similarly, the sloth has a flexible, curved spine. The hoofed mammal, in contrast, has a stiff, straight spine, from which the rib cage and internal organs of the torso are suspended. A deer would be as ungainly in a tree as a sloth is on the ground.

This contrast is mirrored in the teeth. Hoofed mammals have deeply rooted, very hard teeth with ridges of enamel that withstand the toughness of grass. Enamel is the hardest substance a mammal can produce, and, as already mentioned, sloth teeth have no enamel coating. In addition, more than in other mammals, the form and chewing surfaces of the sloths' teeth are sculpted during usage. "Since sloth teeth acquire their individual characteristics through wear, it is very difficult to distinguish the young of one genus from those of another based upon shape or location of dentition" (Naples 1982, 18). In other mammals—especially the grazers—the teeth are preformed with all their crown cusps and ridges before they erupt. The sloth's teeth emerge as simple cones and take on a characteristic form in the course of life.

The sloth is, in this sense, formed from the outside. In a related way we see this tendency in its coloring, which arises not only from hair pigmentation but also through algae from the surroundings. Similarly, its temperature varies with the ambient temperature.

From a different vantage point we can say: incorporating solidity and stability into the skeleton allows a quadruped mammal to live actively within gravitational forces. In giving itself over to gravity, the sloth incorporates inertia. We see inertia in its movements and digestion. The sloth is a bit like the clump of leaves or the alga-covered tree trunk it outwardly resembles.

Drawing In

Active arboreal mammals, like monkeys, have, of course, nothing of the skeletal rigidity of ground-dwelling quadrupeds. They have flexible joints and muscular agility that allow for actively swinging, jumping, and grasping. A sloth lacks the quick and nimble dexterity of monkeys, although it possesses a flexible spinal column (especially in the neck region) and limber fore and hind limbs. A sloth can twist its forelimb in all directions and roll itself into a ball by flexing its vertebral column.

Characteristically, the sloth makes use of this flexibility in its slow movements while feeding and also in protecting itself from a predator by curling up into a ball. The monkey, in contrast, engages in light and springy movements. This leads us to a slightly different way of characterizing the sloth. Its primary gesture is that of pulling in or retracting. It doesn't project actively out into its surroundings.

We can see this tendency in the head. The head is the center of the primary sense organs through which an animal relates to its environment. As we have seen, the eyes and ears are not the sloth's main senses. The outer ears (pinnae) are tiny and hardly visible on the head and the eyes can retract in their sockets. Both of these characteristics reveal externally the muted function of these organs within the whole animal. They also let the head appear as a broadened neck. But this appearance also has a deeper anatomical basis, since the first cervical vertebra (the so-called atlas) is nearly as wide as the widest part of the skull.

The skull itself is rounded and self-contained—superficially resembling a monkey's skull more than a grazing herbivore's (figure 12.5).

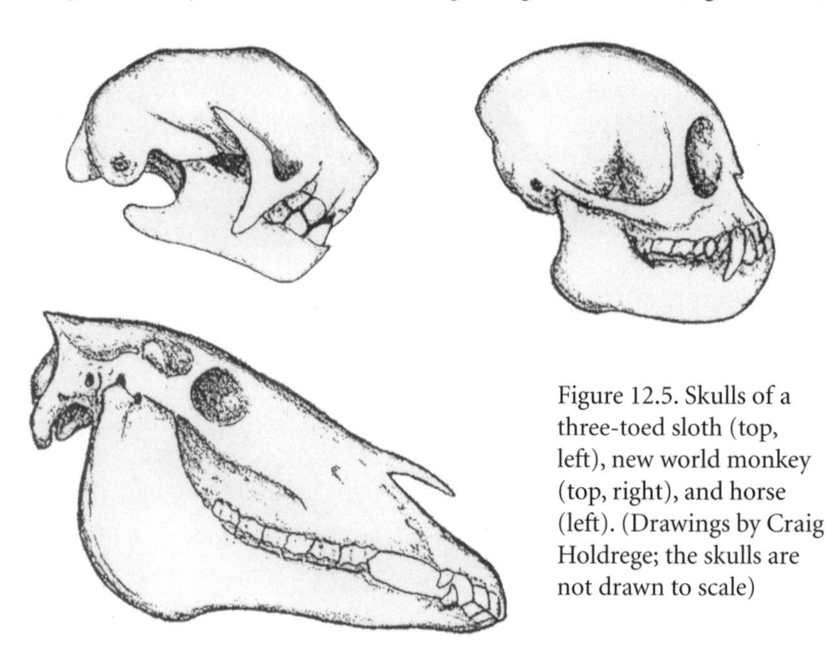

Figure 12.5. Skulls of a three-toed sloth (top, left), new world monkey (top, right), and horse (left). (Drawings by Craig Holdrege; the skulls are not drawn to scale)

Most herbivores have an elongated snout that they use as a limb—standing as they do on all four legs—to reach their food. The sloth's forelimbs have this function, and thus its snout is short. The premaxillary bones—important in forming the elongate mammalian snout—are tiny in the sloth. Moreover, the upper jawbones (maxillae) and the nasal bones are also short in the sloth. The sloth's skull does not project forward.

We have seen that the sense of smell is the sloth's primary sense and that its gesture is to draw in. When we see these facts together with the others, such as the dominance of retractor muscles, then the sloth's special orientation to its surroundings comes more clearly into view.

The Sloth as a Habitat

As if to emphasize its passive, somewhat withdrawn character, the sloth functions as a habitat for myriad organisms. The algae that live in its fur give the pelage a greenish tinge. In addition to the usual ticks and flies that infest the skin and fur of other mammals, a number of sloth-specific moth, beetle, and mite species live on the sloth and are dependent upon it for their development. The sloth moths and beetles live as adults in the sloth's fur. Some species live on the surface, and others inhabit the deeper regions of the fur. They are evidently not parasitic; their source of food is unknown.

When the sloth descends from a tree to defecate and urinate, female moths and beetles fly off the animal and lay their eggs in the sloth's dung. The wings of one moth species break off soon after they inhabit the sloth, making them incapable of flying. Consequently they must crawl off the sloth to reach the dung. The sloth's relatively long period of defecation, which lasts a few minutes, gives the insects the time they need. In this way the slowness of the sloth serves these most "slothful" of sloth moths!

The larvae develop in and feed on the dung (which, you remember, decomposes slowly). The larvae pupate in the dung, and the winged adult moths (or beetles) fly off to inhabit another sloth. Various species of insects and mites inhabit any given sloth, and the numbers of specimens of each species varies greatly, ranging from a few to more than a hundred.

The sloth has been observed grooming its fur. This is typical mammalian behavior and does rid an animal of some of its "pests." From this utilitarian point of view, the sloth's grooming is not very effective. Typically, sloths groom slowly, and sloth moths "may be seen to advance in a wave in front of the moving claws of the forefoot, disturbed, but by no means dislodged from the host" (Waage and Best 1985, 308). Clearly, the measured pace of life, the unique excretory habits, and the consistency of dung allow the sloth to be a unique habitat for such a variety of organisms.

Sensing a Boundary

The expression of pain is a barometer for the way an animal experiences its own body in relation to the environment—the external world is penetrating and harming its biological integrity. Here's an example from a family that kept a sloth at their home in Brazil: "'Sloth burning!' . . . we leap to our feet and run frantically round trying to discover where [the sloth] has fallen asleep. On the kitchen stove? No! On the water heater in the bathroom? No! There he is on top of the floor lamp in the drawing-room, with his bottom touching the big electric bulb! . . . We struggle to get him down, but he clings desperately to his perch, refusing to budge and protesting with many ah-eees against our unwarranted disturbance of his slumbers" (Tirler 1966, 27).

Sloths are reported to "survive injuries that would be deadly within a short time to other mammals" (Grzimek 1975). "I have known a sloth to act normally for a long time after it had received a wound which practically destroyed the heart" (Beebe 1926, 32). These examples show that the sloth does not seem to notice intrusions of its boundaries and continues to live despite them. Its body is not imbued with sensitive reactive presence.

A Further Dimension of Wholeness: The Environment

Where does the sloth end? This seemingly naive question points to a problem and, at the same time, to a task. The problem is the way we think of an organism in relation to its environment. The environment is that with which an animal interacts. Inasmuch as the sloth eats leaves, leaves belong to its environment. In the moment it is interacted with (for example, in feeding, smelling, moving), the environment *is* part of the animal. We could also say, the animal is *part* of its environment. The environment as a functional concept is inseparable from the organism (Riegner 1993). The corpus of an animal with its definite outline—what we call the body—fills a definable volume in space. But the animal's activity carries beyond this corpus. And the environment is part of this activity; without the environment there would be no activity.

It may seem strange to say that the environment is not outside the animal. But this is only because we use spatial terms to describe something functional. Because it is more natural for us to think about the world in the framework of objects, we consider the organism *here* and the environment *there.* But this accounts only for the bodily aspect of the organism, and not its functional and behavioral relations. When we shift our focus from the body as a thing to the body as focal point of activity, then the organism encompasses, firstly, all activities radiating to and from this focal point and, secondly, everything we consider to be outside the organism *before* we change to the functional mode of viewing—leaves, branches, scents, and so on. (We have spoken and will continue to speak of organisms *and* their environments, otherwise we would have to create some new, probably cumbersome, terminology. There is no problem using existing terminology as long as we can see through it to the expanded concept.)

Viewed in this way, organisms actually interpenetrate. Sloth, tree, sloth moth, and algae are all part of one another. We can, therefore, in principle, understand how an ecological community, an ecosystem, and even the whole earth can be considered as further dimensions of organisms. Speaking of the earth as an organism is then no longer merely an analogy, but becomes a reality one has in part begun to grasp—in this case, through the sloth. (And because we have already seen a part, we have also caught a glimpse, in it, of the whole.) It remains the task of a truly holistic or organismic ecology to concretely apply this way of viewing to ecological phenomena.

Is There a Cause of Slothfulness?

In his 1971 compendium on sloths, M. Goffart includes one section entitled "Slothfulness." He describes observations in the field, experimental results, and the hypotheses of scientists concerning the causes of slothfulness. Various possible explanations are brought forth: small heart, slow speed of muscle contraction, low body temperature, low rate of thyroid function, and so forth. He describes the shortcomings of each particular hypothesis and concludes that the "evidence as to the real causes of slothfulness is thus far from complete" (95). If he were writing today, he might include conjectures about genetic mechanisms.

Goffart points out, for example, that the sluggish koala has a constant body temperature of 36 degrees Celsius. Since this is a normal body temperature for mammals, it seems evident that it cannot be causing the koala's sluggishness. Since causes are assumed to be general, he concludes that temperature will also not be the cause of slothfulness in sloths.

Goffart assumes that the causes of slothfulness will one day be found; we are just lacking the necessary information. But this assumption is questionable, and the above example shows, in fact, primarily the limitations of the conceptual framework. In treating aspects of an organism as potential causes, we conceptually lift them out of the organism. Then we think of them affecting things in the organism as though they were not part of it. By so doing we can think in general terms of the factor "body temperature" as a cause, as if separate from the organism.

But every time we carry through this process we realize that our conceptual scheme doesn't fit reality, because we are confronted with mutual relations, all of which express something of the animal as a whole. If we drop this scheme, then it becomes interesting that body temperature evidently means two very different things in the koala and the sloth. Instead of looking for genetic or physiological causes that we assume have general validity, we look at the unique expression of physiological facts in the given context. We take the unique integrity of each animal seriously.

It is second nature for a scientist to inquire after the causes of what is under investigation. Some would even say that this is the task of science. But in the context of organisms this method alone is not adequate. Putting it a bit radically, biologists would do well to eradicate the term "cause" from their vocabulary and use the more modest and open term "condition." What genetic, physiological, behavioral, and ecological studies can show is how aspects of an organism provide mutual and changing conditions for one another. This knowledge is extremely valuable as long as we don't separate it from the organism as a whole. In fact, it can be the gateway to understanding the organism as an integrated whole.

Encircling the Unspeakable: The Animal as a Whole

Let's return to the statements quoted at the beginning of this chapter: George Louis Leclerc, Comte de Buffon, was a well-known eighteenth-century French scientist. He studied many animals, among them the sloth, about which he said: "one more defect and they could not have existed" (quoted in Beebe 1926). He considered the sloth's remarkable characteristics to be defects. And they are, if you take the point of view of a horse, eagle, jaguar, or human being. But as naturalist William Beebe countered, "a sloth in Paris would doubtless fulfill the prophecy of the French scientist, but on the other hand, Buffon clinging upside down to a branch of a tree in the jungle would expire even sooner" (Beebe 1926, 13).

Buffon takes a standpoint outside the animal. We have followed Goethe's suggestion and tried to view the sloth on its own turf. Goethe wrote: "Hence we conceive of the individual animal as a small world, existing for its own sake, by its own means. Every creature is its own reason to be. All its parts have a direct effect on one another, a relationship to one another, thereby constantly renewing the circle of life; thus we are justified in considering every animal physiologically perfect" (Goethe 1995, 121).

We have made use of comparison, but not to describe what the sloth "should" have in order to be a reasonable animal. The animals described by way of comparison shed light on the sloth, allowing its uniqueness to stand out all the more perceptibly. When Goethe calls an animal "perfect," he means that each animal has its own unique way of being—its specific integrity that we can try to understand. But this is no simple matter. Goethe recognized that "to express the being of a thing is a fruitless undertaking. We perceive effects and a complete natural history of these effects at best encircles the being of a thing. We labor in vain to describe a person's character, but when we draw together actions and deeds, a picture of character will emerge" (1995, 121, translation modified by Craig Holdrege). In trying to paint a picture of the sloth, we have discussed many details, because through them the whole lights up. Henri Bortoft puts it well when he says, "The way to the whole is into and through the parts. The whole is nowhere to be encountered except in the midst of the parts" (1996, 12).

This emergent picture of the whole does not and cannot encompass the totality of its characteristics. One can always discover new details. It is not a matter of striving for totality, but rather for wholeness. Our understanding hinges on our ability to overcome the isolation of separate facts and to begin to fathom the animal as a whole, integrated organism. The whole is elusive, and yet, at every moment, potentially standing before the mind's eye. When we begin to see how all the facets of the animal are related to one another, then it comes alive for us. Or, putting it a bit differently, the animal begins to express something of its life in us. Every detail can begin to speak "sloth," not as a name, but as a qualitative concept to which a definition can do little justice.

We have tried to describe the sloth in a way that allows us to catch glimpses of its wholeness. We can now refer to such characteristics as slowness, inertia, blending in with the environment, receding or pulling in and not actively projecting outward. Each expression is a different way of pointing to the same coherent whole. Taken alone, as abstract concepts or definitions, they are empty. They are real only inasmuch as they light up within the description or perception of the animal's characteristics. But they are not things like a bone or an eye. They are, in context, vibrant concepts that reveal the animal's unique way of being.

Let's return to the sloth, high in the crown of a rain forest tree, hanging from or nestled on a branch. In its outer aspect, it blends in with its environment. There are no sudden or loud movements. The sloth's green-tinged, mottled brown coat lets it optically recede into the wood and foliage of its surroundings. And like the tree bark, the sloth's fur is teeming with insect life. The sloth's body temperature rises and sinks with the ambient temperature.

The round form of its head is the anatomical image of the way in which the sloth does not actively project into its environment. There are no large, movable, reactive outer ears, and the eyes are rarely, if ever, moved. The sloth has no protruding snout. It draws the scents of the environment, especially of the leaves it feeds upon, into its nose. But much of the day the sloth is curled up, unaware of the world around it. Even when awake, the sloth seems not to live as intensely in its body as other mammals, being quite insensitive to pain.

The sloth does not carry its own weight; rather, it clings to an outer support. Its skeletal system is not characterized by stability, but by looseness. This laxity allows the sloth to adopt positions that would be contortions in other animals. The sloth makes mostly steady pulling movements with its long limbs, a capacity based on the dominance of retractor muscles.

The sloth develops slowly in the womb and has a long, slow life. It moves slowly through the crowns, feeding on the leaves that surround it from all sides, bathing, as it were, in its food source. The leaves pass through the animal at an almost imperceptibly slow rate. The sloth's stomach is always filled with partially digested leaves. Even its dung disappears slowly, despite the warm and humid rain forest climate that normally accelerates decomposition processes.

The sloth brings slowness into the world.

Part IV

Science Evolving

Chapter 13

The Language of Nature

• I •

To judge from some of the ancient creation narratives, the world arose as a visible manifestation of speech. "In the beginning was the Word." First there was formlessness and chaos, and then the divine voice flashed forth like lightning in the darkness. "And God said, Let there be light: and there was light." The world began to assume visible, comprehensible form.

Whatever we may now think of the old visions of creation, we can remain sure of one thing: without the speaking of the Word—without language—we would have no science, with its striking power to illuminate the world. This observation may seem trite; no one will deny that we must use words in order to achieve and record our scientific understanding, or to pass it on to future generations. But once we stop to reflect upon the fact that science is always a science of speech, a remarkable thing begins to happen. We find ourselves transported to a richly expressive realm of scientific meaning that is as far removed from cramped, traditional notions of science as the first day of creation was from the primeval chaos.

The truths capable of revolutionizing our understanding can sometimes be so close to us that we fail to notice them. So it is with science and language. The crucial point is easy to miss: it's not that we humans just happen to need words in order to talk scientifically about a world that in its own right has nothing to do with language. Rather, it's that our need for words testifies to the word-like nature of the world we are talking about.

I (Steve Talbott) realize that this last statement will provoke surprise and skepticism in many readers of our day. And yet, as long as there has been science, leading scientists have routinely referred to the "language of nature"—not, perhaps, with a clear notion of their own meaning, but with an evident comfort and sense of rightness about the usage. There is good reason for this. After all, the whole point of our language, our speaking, is to characterize something other than our own speech. When we say "atom" or "energy" or "mass," we are speaking *about* something. We seek to elucidate an aspect of the world. To the extent that the meaning of our scientific descriptions is not at the same time the meaning of the world, the descriptions fail as science. As scientists we are always trying to speak faithfully the language of nature.

In slightly different terms: the world is in some sense a text waiting to be deciphered—which is why we can in fact decipher it into a scientific description. As with any text, we expect the world-text to make *sense,* to hold together conceptually, to speak consistently, to justify itself to our reason. These are demands we can bring only to whatever is word-like.

The intimate relation between the meaning of our words and the meaning we find in the world may be so obvious as to seem almost trivial, yet its implications are so profound as to have mostly escaped the notice of working scientists. If we took the fact of the world's speech seriously—the world *speaks!*—there would be none of the usual talk about a mechanistic and deterministic science, about a cold, soulless universe, or about an unavoidable conflict between science and the spirit. Confronting the many voices of nature, we would inquire about their individual qualities and character, we would look for the direction of their expressive striving, and we would struggle to grasp the aesthetic unity of their various utterances—all of which is to say: we would listen for their *meanings.* The necessity for such inquiry is implicit in a world that speaks and also in the scientist's employment of speech to translate the world text. What I wish to suggest is that, by turning a deaf ear to a resonant world and even to our own speech, we underwrite many of the limitations and contradictions of the science we have today.

As for what I mean by speech and word-like, I hope this will emerge with greater clarity over the course of this chapter. Suffice it to say for now that everything word-like presents itself as a perceptible exterior bearing an inner and partly conceptual meaning (Barfield 1977).* The

* Purely conceptual content—thought without words, to whatever degree we are capable of such thought—is taken to be word-like in a higher sense than the perceptibly embodied concept (or word). This chapter could have spoken about the "concept-likeness" of the world, but given the intimate connection of thought and language, together with the modern inattention to pure thought, it seemed more immediately understandable to take language as the starting point.

meaning of words is never found in the mechanisms or physical causes of their production. No chemical analysis of the ink on the page, no physical analysis of the act of writing, or of the speech apparatus and the air-forms it produces, can by itself give us the inner content of the words. That's because meanings and concepts are immaterial; they are not tangible or otherwise sense-perceptible things, which is what I will indicate by saying they are *inner.* One could also say: meanings are always contents or expressions of consciousness. When we find a written text meaningful, we rightly assume that it is the product of conscious activity, and we ourselves can know the meaning only in so far as it lights up in our own consciousness.*

Most of us have had experience successfully reading the meaning of texts and hearing the meaning of speech, and therefore in this practical sense we already understand words as bearers of meaning. And just as we find our own speech vitally supplemented by physical gestures of every sort—gestures that are themselves outward bearers of inner meaning—so, too, all of nature presents us with word-like gestures. What I will be asking of you as a reader is to bring all this meaning—your experience of it, and not a definition of it—as vividly as possible to consciousness.

Drawing a Blank

Fish swim, and their capacity for swimming makes no sense without water. Birds fly; their entire structure and functioning testify to the sea of air in which they live. And we humans speak; we navigate a sea of meaning. As the bird and fish necessarily evolved in relation to their environment, so did we. Our speaking was made possible by the world's meaning. This meaning is no more an arbitrary and subjective invention of our own than the ocean is an arbitrary and subjective invention of the fish.

No one will deny that we *experience* meaning everywhere in nature. To sit in a quiet glade with the sun streaming through the trees; to endure the shattering power of a fierce thunderstorm; to enjoy the early greening of spring or the warm, rich colors of autumn; to stand beside a quiet pond or the rapids of a stream; to climb toward the summit of a high peak; to watch the unfolding drama of a sunset; to lie down and gaze up at the stars—every setting we encounter comes to its own

* "Consciousness" is used in the broad sense, so as to include those so-called subconscious contents not yet known, or fully known, at the self-conscious center of our being.

meaningful and distinct expression within us. Everything speaks an inner language.

But our long-standing habit is to write this experience off as something wholly manufactured within ourselves—the speech, we are inclined to say, is our own, not nature's; subjective, not objective. And since whatever lacks objective value hardly seems worth bothering about in our quest for an understanding of nature, we have little incentive to attend to our experience of the stream or storm and even less to discipline this attention so as to discover scientific value in it. As a result, the experience really does fade into a kind of subjective vagueness, and increasingly we find ourselves drawing a slightly disturbing blank whenever we do try to appreciate the natural world in its own, qualitative terms. Jan Hendrik van den Berg presumably had something like this blank in mind when he wrote:

> Many of the people who, on their traditional trip to the Alps, ecstatically gaze at the snow on the mountain tops and at the azure of the transparent distance, do so out of a sense of duty. . . . It is simply not permissible to sigh at the vision of the great views and to wonder, for everyone to hear, whether it was really worth the trouble. And yet the question would be fully justified; all one has to do is see the sweating and sunburned crowd, after it has streamed out of the train or the bus, plunge with resignation into the recommended beauty of the landscape to know that for a great many the trouble is greater than the enjoyment. (van den Berg 1975)

Few of us can altogether disclaim the experience of those tourists. Even many who are capable of more refined attention to nature will, I suspect, sympathize with my own plight: when I venture into the wild, something in me recognizes many "stunning" and "enchanting" things, and yet these things don't speak to me with any clarity. I am continually drawn to them, sensing that they *should* speak to me with a force much greater than I am capable of receiving, but I am largely dispossessed of whatever understanding of their language humanity may once have had.

As for science, the problem of incomprehension seems to disappear only because nature's speaking is more or less explicitly disavowed and therefore not attended to. One doesn't even bother to get out of the bus. It's enough to mount some instruments at the windows so that they can "observe" nature for us. This habit of inattention was asserted as a matter of principle almost from the beginning, when Galileo banned qualities from his science. Tastes, colors, and odors, he claimed, are "mere names" that "reside only in the consciousness." External reality manifests nothing but shape, number, and movement, which, it happens, lend themselves to mathematical treatment (Galilei 1957, 274–77). To rid science of qualities in this way, preferring quantitative demonstrations alone, was to push along the straightest path toward the elimination of meaning from science.

If, as I have suggested, nature is a speaking and science is one sort of translation of this speaking, then the decision to turn a qualitatively deaf ear to nature's voice ought to be writ large in our scientific language. And so it is. In fact, language can show us with striking vividness the character of the blank that nature has become for us.

How to Ignore Meaning

In my primary school days it was still the common (if widely resented) practice for students to diagram sentences. The diagram offered a way to display as clearly as possible the grammatical structure of our language.* Given the sentence, "William hit me," you could show the relation between the subject ("William"), verb ("hit"), and direct object ("me") with the aid of a few lines:

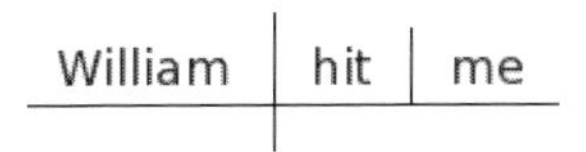

Words that modify other words are placed on slanting lines beneath the words they modify, so that the sentence, "The large, black dog bit the postman," yields this diagram:

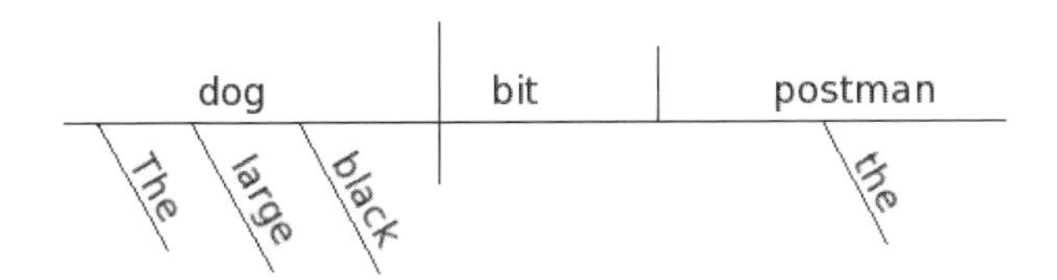

* These traditional grammatical categories, however, are extremely loose, and such diagrams are poor cousins of the much more rigorous and formal ongoing effort to articulate mathematically strict grammars.

Prepositional phrases ("in the leg") can be diagrammed to show the relation between the preposition ("in") and the object of the preposition ("leg"):

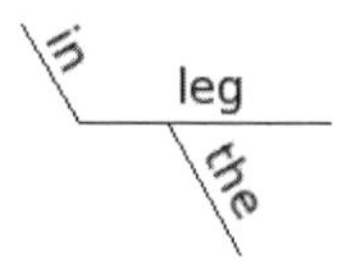

And since the whole prepositional phrase modifies another word ("bit" in the sentence below), we can diagram the complete sentence, "The large, black dog bit the postman in the leg," in the following manner:

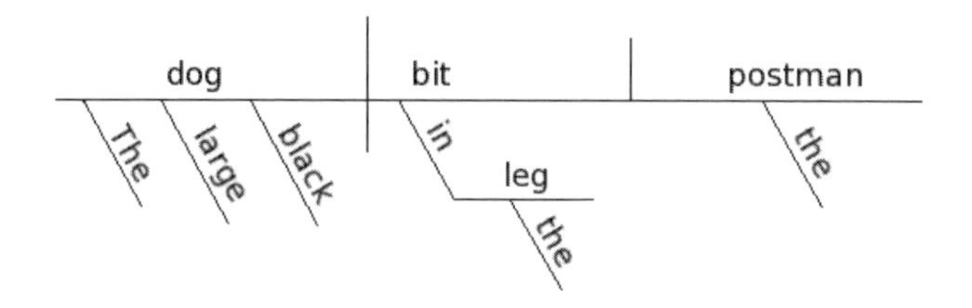

Or consider the following sentence, in which I have added a few new features, including a conjunction linking separate clauses—and, before reading further, try to make a guess of your own about the grammatical correctness of my diagram:

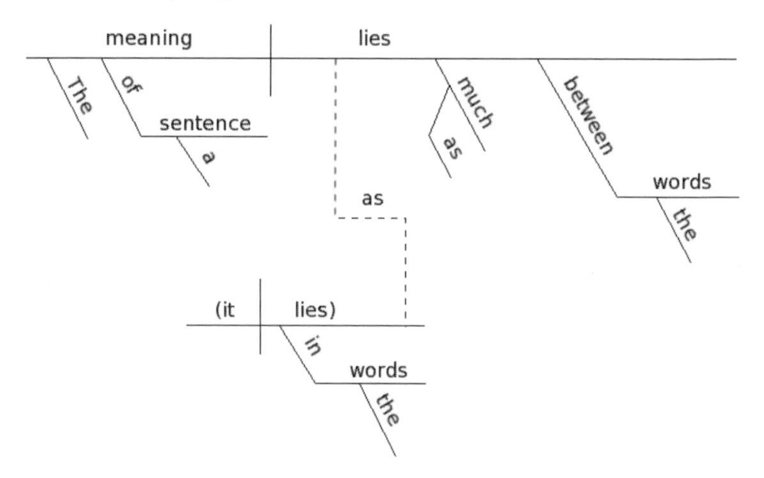

"The meaning of a sentence lies as much between the words as in the words."

Unlike most of my classmates, I loved to diagram sentences. There was pleasure in grasping a relatively straightforward, unequivocal truth about words at the level of their structural arrangement. What I didn't notice at the time was that in the act of tracing the diagrammatic structure of the sentences, I was not attending in any full way to their meaning. Ask yourself: in trying to judge the structural correctness of the sentence diagram above, did you find yourself deeply engaged with the meaning of the sentence (which certainly deserves to be puzzled over a good bit)? Perhaps not. And if you did, you will certainly recall that your attempt to arrive at the correct diagram was an activity more or less separate from your ruminations about possible meanings.

When you are diagramming sentences, the concrete and particular disappears into the abstract and general. The main thing you want to know about each word—such as "black" in "The large, black dog bit the postman in the leg"—is what grammatical category it belongs to. The meaning of the word scarcely matters; it could be "brown" or "fierce" or "crippled"—or even something with little sense, such as "prayerful" or "zodiacal"—and this would neither change the diagrammatic structure nor affect its correctness.

When we are diagramming a sentence, our understanding of it contracts into something precise and demonstrably correct, but our ease and precision of judgment is obtained by eviscerating the sentence of its full and particular content. Our attention is narrowed from the meaning of the words to a highly abstract feature of them. For example, all descriptive words of a certain sort become merely the same thing—"adjectives." Words of another sort become nothing but "prepositions." Once we have learned the rules for diagramming sentences, we can obtain correct diagrams almost while "running on automatic"; but the meaning of the diagrammed sentences—especially if they are at all profound—is far from automatically fathomable, and it would often be rather arrogant to say, "Here is *the correct* meaning of this sentence" in the way you might claim a correct diagram.

The operation of abstraction is perfectly legitimate and valuable in its place, but to forget what it removes from consideration—or that by itself it leads us progressively toward an emptying of meaning—is not helpful when we want to understand the words, or the world the words are meant to illuminate. When we crucify the world-text upon a scaffold of grammatical logic, the resulting corpse presents its own fascinations, but these are not the fascinations of the original meaning; they are only a shadow of it.

On Being Wonderfully Precise about Practically Nothing

Diagramming sentences is only one of the ways we can reduce a full-fleshed text to a skeleton that we can nail down with greater exactitude. We move in the same direction when we reconceive a text as information (in the technical sense given by communication theory), as a logical or mathematical structure, or as an algorithm. But every effort of this sort brings us up against a general truth of overwhelming significance for our science and for any resolution of the perennial conflict between science and the humanities. Warren Weaver was getting at this truth when, in his introduction to *The Mathematical Theory of Communication,* he wrote that strict, mathematically defined information, on the one hand, and meaning, on the other, appear "subject to some joint restriction that condemns a person to the sacrifice of the one as he insists on having much of the other" (Shannon and Weaver 1963, 28). More generally: in all use of language, precisely delimited, unqualified understanding tends to be purchased at the cost of content, or fullness of meaning, and fullness of meaning tends to be purchased at the cost of precise, unqualified definition.*

When we drive language as far as we can toward the pole of precision and definitive, yes-or-no certainty, we arrive at formalisms such as mathematics, grammar, and logic. In the case of pure logic, the withdrawal from meaning or content is so extreme that the logician conscientiously refuses to speak of the "truth" of his logical propositions. Instead he refers to their "validity"—their internal consistency without reference to any content of the world. And so Bertrand Russell, one of the preeminent logicians of the twentieth century, once remarked of mathematical logic that it "may be defined as the subject in which we never know what we are talking about" (Russell 1981, 59–60). Einstein expressed the same thought this way:

* Owen Barfield briefly refers to the polar relation between accuracy and fullness of meaning in *Speaker's Meaning* (1967, 35–39). But his earlier work, *Poetic Diction* (1973, first published in 1928), can be read as a book-length study of these "polar contraries," without, however, referring to them as such. And his rather difficult text, *What Coleridge Thought* (1971), deals with polarity as a central theme. I (Steve Talbott) am deeply indebted to Barfield for my understanding of the notion of polarity, and therefore for the entire content of this chapter.

> The skeptic will say: "It may well be true that this system of equations is reasonable from a logical standpoint. But this does not prove that it corresponds to nature." You are right, dear skeptic. Experience alone can decide on truth. (1954, 355)

> Pure logical thinking cannot yield us any knowledge of the empirical world; all knowledge of reality starts from experience and ends in it. Propositions arrived at by purely logical means are completely empty as regards reality. (1954, 271)

Mathematics and logic as such are not about a *what*—not about the world's actual and particular phenomena—but rather provide empty templates for thinking about the world in a certain way. *But we still have to do the thinking,* and we cannot bring the world into this thinking while remaining solely within the self-contained and reassuring purity of the templates. The world breaks every fixed template into which we try to pour it. Referring back to our diagram: if we can substitute one adjective for another without affecting the correctness of our structure, then we have to acknowledge that the diagram fails us badly as an adequate explication of our speech; it cannot distinguish between any of our meanings. Words hopelessly overflow the expressive power of the diagram.

Despite the fact that formally manipulable quantities and the content of observation have very different character, they are intimately woven together by the scientist, even if the weaving does not often receive critical attention. This is both necessary and proper. But it is vital to understand the differing tendencies of the fabric's warp and woof, and to recognize the unbalanced extreme to which the prevailing bias toward the quantitative and formal will lead us if it is not countered by something working at cross-purposes with it.

At least in the case of sentence diagrams we still have the meaningful text alongside the abstract grammatical structure indicated by the lines of the diagram. We can refer back to this text and relate our abstract construction to it at any time. We can re-enflesh the formal skeleton. Often in the "hardest" sciences the world-text—the phenomenon we began by trying to understand—disappears entirely behind multiple layers of theoretical construction, as when a rainbow becomes light rays and raindrops, which in turn become photons, molecules, and subatomic particles, which in turn give way to the clean statistical abstractions of quantum mechanics.* Not only are the mathematical threads in our tapestry of cognition by far the most highly regarded, but "observation" has come increasingly to consist of the gathering of quantitative *data,* so that our tapestry begins to look like all warp and no woof.

Science deeply colors our experience of the world today. Is it any wonder that van den Berg's tourists should draw an unhappy blank when gazing upon the Alps? We have learned through long habit to discount the speaking content of nature as a vagary of our own detached consciousness, superfluous in relation to the mathematical grammar that, we are sure, must be the potent, if incomprehensible, essence of what lies all around us.

It would be healthier if we could begin questioning our scientific inheritance. It is hardly impertinent to point out that if, in the interest of precision, we narrow our technical language down to an empty formalism, then we are not *discovering* the world to be meaningless; we are *insisting* that it be meaningless. There can be discomfort and threat in any confrontation with a speaking presence, and perhaps we should open ourselves to the possibility that much of our satisfaction in the unqualified rigor and precision of our science is really the satisfaction of curling up within the secure refuge of speechless quantity and logic, without having to venture too far out into the complex, soul-gripping presentations of the phenomenal world.

Minor Refinements or Wholesale Revision?

The conflict between the official banishment of meaningful, qualitative language from hard science on the one hand, and the inevitable reliance upon it on the other, has led to a strange sort of schizophrenia. Physi-

* James Lovelock writes in *Gaia*:

> Unfortunately, most scientists live their lives in cities and have little or no contact with the natural world. Their models of the Earth are built in universities or institutions where there is all the talent and the hardware necessary, but what tends to be missing is that vital ingredient, information gathered first-hand in the real world. In these circumstances it is a natural temptation to assume that the information contained in scientific books and papers is adequate, and that if some of it does not fit the model then the facts must be wrong. From that point, the fatal step of selecting only data which fit the model is all too easy, and soon we have built an image not of a real world, which might be Gaia, but of that obsessive delusion, Galatea, Pygmalion's fair statue.

Lovelock adds that personal contact between the model builders working in city-based institutions or universities and those relatively few who explore the world "is rare

cists today employ mathematical constructions so neatly cohering and so universal in their logical coverage that some researchers, such as string theorist Brian Greene, have wondered aloud whether they are closing in on a finished "theory of everything," while others have publicly debated "the end of science." Yet many of these same physicists are driven by their work toward a kind of rootless, unrestrained, almost childish speculation about the nature of things. I will offer illustrations in a moment. But to see what is going on here, first consider a familiar case.

At a time when scientists were learning to observe and measure very high velocities, Einstein was led to the startling and unexpected theory of relativity. But if, in a scientific gathering today, you were to cite this theory as an example of the susceptibility of science to wholesale and fundamental revision, you can be sure that some of your listeners would respond by saying, "Einstein did not prove Newton's prevailing formulas to be wrong; he merely showed them to be approximations in need of further refinement—extremely minor refinement under those conditions Newton was able to survey. Relativity did not so much negate Newton as confirm his results and extend them to cover more extreme conditions. The 'correction' is trivial under most normal circumstances."

And this is true! At least, it remains true as long as we reside within the narrow, quantitative terms of our scientific laws or "grammar of nature." But this is to ignore what was in fact a revolution in our understanding. The revolution becomes apparent as soon as we try to hear the meanings that alone enable our grammatical refinements to speak of the world. Physicist David Bohm reminds us that "while the laws of relativity and quantum theory do in fact lead under special conditions to small corrections to those of Newtonian mechanics, they lead more generally, as is well known, to qualitatively new results of enormous significance, results that are not contained in Newtonian mechanics at all" (1971, 133). Likewise, referring to relativistic effects upon mass, Richard Feynman writes: "*philosophically we are completely wrong* with the approximate laws [such as Newton's]. Our entire picture of the world has to be altered even though the mass changes only by a little bit" (Feynman et al. 1963, 1–2, emphasis in original).

One wonders why he says "philosophically" and not "scientifically." Is science really incapable of giving us a "picture of the world," so that

and information passes through the terse limited phraseology of scientific papers, where subtle, qualifying observations cannot be included along with the data" (Lovelock 1987, 136–37). See also Holdrege 2006.

this picture must be left to the philosophers? In any case, if we want to understand the *world,* and not merely define more accurately certain regularities of its grammar, then we must grant that Einstein's was an altogether different world from Newton's, requiring a new way of conceiving the fundamental elements of space and time. To say that the changes Einstein introduced to our scientific understanding were "minor" is like taking the sentence, "The large, black dog bit the postman in the leg," changing "black" to "invisible," and then asserting that the change in the sentence is minor—or indeed nonexistent—since the diagrammatic structure remains the same.

A science that can deceive itself in this way is a science that can all too easily say, "Our knowledge leaves no room for the human 'soul' or 'spirit.'" And it's true that the identification of science with empty formal structure leaves no room for soul and spirit. But it leaves no room for anything else, either. One can agree only with the first half of physicist Steven Weinberg's remark in *Dreams of a Final Theory*: "The reductionist worldview *is* chilling and impersonal. It has to be accepted as it is, not because we like it, but because that is the way the world works" (1992, 53).

Weinberg should rather have said, ". . . because that is the way my preferred language works—the only language I wish to accept as scientific." This language can seem impersonal and chilling only because we have reduced it to a grammar that necessarily ignores whatever understanding we might gain of the world's meaningful content.*

Deep Math and High Speculation

Once you have sacrificed meaning in order to arrive at your well-behaved grammatical abstractions, there is no way to recover the lost meaning from the abstractions alone. This is why physicists today, despite sharing an admirably exact mathematical grasp of the "fundamental laws of the universe," give us the most amazingly different worlds when they try to imagine the reality from which these laws were abstracted—the reality that actually embodies the laws and lends them meaning.

* The usual way to describe the relation between scientific laws and observed phenomena is to say that, in order to explain the phenomena, we need not only laws but also a specification of the "initial" (or "boundary") conditions. But this scarcely brings out the crucial requirement. We do not need merely to know "where things start," as if these things could be taken for granted; rather, we must first gain the things themselves as real content. And this descriptive task turns out to be the central work of science (see "To Explain or Portray?" below).

So it is that the journal *Scientific American* can advertise one of its publications by asking, "Is there a copy of you in another universe, reading this sentence?" And, we are assured, "the most popular cosmological model today suggests that the answer is yes." The advertisement goes on, however, to note that physicists disagree in how they understand this notion of parallel universes, with some seeing the different realms as "wildly dissimilar," displaying wholly different laws, and others seeing them as near-copies of each other. Such speculation leads the *Scientific American* writer in the familiar direction taken by so many scientists when they try to explain themselves to a popular audience—namely, toward language that is almost mystical: there is, the advertisement tells us, "another possible plane of reality (one where you are most definitely not alone)."

Don't feel badly if you're mystified about this other plane of reality; so, it seems, are the scientists who sell books by employing such language. Their divergent speculations would make the most levitated medieval metaphysician blush. These speculations go far beyond parallel universes and tend to arise whenever researchers try to explain what sort of world their equations are about. Are nature's laws founded upon absolute randomness? Can time flow backward? Are there wormholes that take a shortcut through spacetime, linking two different times? Is there a shadow universe sharing gravity, but no other forces, with our own universe? Can we know the "real" world at all? Does observation create reality? Does consciousness create reality?

The ground under our feet becomes still more unstable when we consider how even the most basic terms of routine scientific explanation are more or less blank. It was no high confession, but a simple recognition of the obvious, when Richard Feynman remarked that "we have no knowledge of what energy *is*" (Feynman et al. 1963, 4-2). Much the same is true of all the basic terms of science referring to the phenomenal world: gravity, light, heat, space, time, and so on. The language and methods of physics simply don't aim at discovering a meaning or content for these terms.

The theory of everything, it seems, comes perilously close to being a theory of nothing, or, at least, nothing very meaningfully understood—exactly what you'd expect when the theory's glory and substance are taken to lie in its purely grammatical or formal lawfulness. If the physicist's speculations about the nature of the universe sometimes seem bizarrely untethered, it's because *there is not enough reality in the parameters of this science to constrain interpretation*—which is one example of the general fact that there is not enough reality in a formal grammar, or in a formalism of any sort, to constrain our understanding of the content expressing itself through the formalism. If we employ a reduced scientific language inadequate to express the world's reality, we will have a science with fantastic and unstable content reflecting undisciplined fancy more than reality.

And this science, dominated in its meaningful aspects by untethered human fancy, is the same science, so we are continually told, that has displaced the human being from his cherished place at the center of the world!

On Perceiving the World as a Machine

There is, however, one reality principle in the hard sciences, and it rules with a vengeance. It is found in the uncompromising (and perfectly healthy, in its place) demand for devices that actually work. What the researcher proposes does not become a part of science until it leads to an experimental apparatus that suffers predictable change under a specified set of circumstances. This technological imperative, with its useful and often striking consequences for our daily life, accounts for much of the popular conviction that science *must* have succeeded in connecting us to reality.

And so it must in one way or another. Our science brings us very real manipulative skills. But the skills enabling us to manipulate a thing are not necessarily the skills yielding deep insight into its nature. In fact, in a world of speech and expression (consider your relations to family and friends), manipulation tends to work directly against understanding. In concerning ourselves with the mechanistic logic we can lay bare in an object, we are throwing a veil over its distinctive expressive character. The following reflection may help to clarify the point.

If you wanted to create a manageable, bounded, relatively self-contained realm embodying your conviction that the world is driven and controlled by a kind of formal necessity—by a pure structure of logic—you could hardly do better than to invent the computer. The entire history of technology has converged upon this apotheosis of mechanistic thought, often referred to by theorists as a "logic machine." Strikingly, the machine's program logic is now taken to *be* the machine, or at least to be what really counts in it. This is all too natural, given that "today's computers have been designed to follow strictly a formalism imposed independent of their physical implementation. The properties of the materials that implement the computation are hidden by careful engineering" (Zauner and Shapiro 2006). And so the same, high-level computational behavior can be designed into devices of radically, almost unrecognizably, different physical nature. This physical nature begins to seem irrelevant. It is not for nothing that computer scientists, preoccupied with their pristine algorithmic structures, often refer disparagingly to the clumsy and recalcitrant "world of atoms" in contrast to the light, lucid, and manageable "world of logical bits."

And now it is this machine, its externally imposed formal purity unsullied by the peculiarities of its material embodiment—a machine whose admired logic therefore gives us virtually no understanding of the physical device we marry the logic to—that has become our reigning model for understanding the physical world. We imagine the world's lawfulness to stand in the same relation to the world as software stands in relation to the computer—but this is a relation in which the lawfulness can tell us almost nothing about the intrinsic character of physical reality!

Along this path the way is open for an ever more complete withdrawal from the world's self-expression. Whatever the wonders we have produced within the closed system of technology, they do not testify to the disciplining of our understanding by physical reality except in a highly impoverished way. The magic of the digital machine is that by squinting at it and looking in just the right way, we can drop the material device from view altogether and see only the clean, universal, eternal pattern of lawfully articulated logical bits that we ourselves have impressed upon the machine. This logic certainly does not picture for us the inherent lawfulness of copper, silicon, glass, and all the rest. Despite this, we are ever more inclined to view the natural world through the mental grid (or chain-link fence) constituted by our logic-machine-ideal, and we thereby reinforce our impossible desire for a universal grammar of nature that somehow explains and determines everything that happens (Talbott 2003b and 2004).

Of course, it requires only a little spilled coffee to remind us that the materials of the computer have their own substance and presence and sometimes maddening behavior not at all accounted for by the light and lucid "governing" laws we have programmed into their physical structure.

Descent toward Primitive Animism?

The displacement of meaning by our grammatical fixation helps us to understand the curious ambiguity in our modern sense of alienation from the world. On the one hand, we imagine a kind of iron-clad necessity imposed by universal physical laws. But because these laws are helpless to determine the world's actual content, our sense of deterministic enslavement is hardly absolute. There is very much an opposite feeling: the typical human complaint in the scientific era has been one of *meaninglessness,* which is a kind of hopeless nondeterminism. That is, the scientific account of the world lacks enough *significant* order, enough pattern and coherence of the speaking sort, enough sense and intention—in sum, enough textual meaning—to provide a context for our own meaningful existence. The problem is not so much that we are cogs in some inexorable machine suborning us to its own purposes, as that our science would allow this machine no purpose at all, no definite character. And so we become lost atoms moving senselessly in the void.

Nobel Prize–winning physicist Steven Weinberg gives unwitting expression to the complex nature of our alienation when he writes that we are the "more-or-less farcical outcome of a chain of accidents reaching back to the first three minutes [after the Big Bang]" and that we are all "just a tiny part of an overwhelmingly hostile universe. . . . The more the universe seems comprehensible, the more it also seems pointless" (1977, 154–55).

Yes, the more we reduce our comprehension of the universe to a mere grammar, the more it seems pointless. But we can't really have a purely grammatical—an altogether empty or pointless—understanding of anything; we cannot have understanding without a content that somehow speaks. When we try, we end up supplying our own content, however crude and unrecognized. This is why Weinberg, despite his belief in an explanatory lawfulness utterly devoid of meaning, naively ensouls this lawfulness with his own meaning: the universe by his account is *farcical* and *hostile*—which is a far cry from being pointless.

Because Weinberg is not actually *looking* at the world's expressive qualities, his assumptions about their character are little more than a kind of animism in scientific dress; his inhospitable animating spirits of farce and hostility reek more of sour professor than of genuine demon.

Those who do look at the world may see elements of farce or hostility in limited contexts, but they certainly see a great deal more.

• II •

The emptiness of scientific language, *just so far as* it fulfills the reigning quantitative and logical ideal, is scarcely open to dispute. It has been recognized, as we have seen, by prominent scientists and philosophers. If you still want to declare the world cold and impersonal, indifferent to human hopes and feelings, relentless and implacable in its mindless obedience to physical necessity—well, that is certainly your privilege. But the one ground least available for your contention is the ground where you celebrate the mathematical precision, certainty, determinacy, and universality of scientific laws. At least, it remains unavailable until you elucidate a path from empty formalism to a revelatory description of the world, and then demonstrate what your precision, certainty, determinacy, and universality mean for this enfleshed world. This in turn will require coming to terms with a paradox that should by now ring familiar: "As far as the propositions of mathematics refer to reality, they are not certain; as far as they are certain, they do not refer to reality" (Einstein 1954, 233).

The amazing thing is how little this puzzle has been taken up by working scientists, especially in the hard sciences. If, as embryologist Lewis Wolpert suggests (1992, 121), "all science aspires to be like physics, and physics aspires to be like mathematics," then, at the very least, we might want to inquire about the adequacy of our aspirations. One physicist who does refer to the problem is Richard Feynman. Alluding to the same theme we have traced here, he reminds the mathematically inclined physicist of the necessity for a step beyond formalism toward real-world meaning:

> Mathematicians are only dealing with the structure of reasoning, and they do not really care what they are talking about. They do not even need to *know* what they are talking about, or, as they themselves say, whether what they say is true. . . . But the physicist has meaning to all his phrases. That is a very important thing that a lot of people who come to physics by way of mathematics do not appreciate. . . . In physics you have to have an understanding of the connection of words with the real world. It is necessary at the end to translate what you have figured out into English, into the world, into the blocks of copper and glass that you are going to do the experiments with. (1967, 55–56, emphasis in original)

But this should not be taken simplistically. It cannot be merely a matter of translating from the language of pure mathematics to the meaning of the physicist because, as Feynman has just acknowledged, there is nothing we can say mathematics is *about*—no content available for translation. Before you can "translate what you have figured out," you must have figured something out—something more than mathematical, having to do with the presence and character of an observable content. This content finds its way into our thinking by processes distinct from the abstract ruminations of the pure mathematician. A formalism itself cannot direct us to any specific content capable of embodying the formalism.

How then *do* we find the content of our science? The ease with which this question has been ignored stands as one of the most stunning features of our science-committed culture. Science historian E. J. Dijksterhuis, describing the shift away from medieval thought during the scientific revolution, tells us that "'substantial' thinking, which inquired about the true nature of things, had to be exchanged for 'functional' thinking, which wanted to ascertain the behaviour of things in their interdependence." For this purpose, "the treatment of natural phenomena in words had to be abandoned in favour of a mathematical formulation of the relations observed between them" (1961, 501). Dijksterhuis seems to find nothing at all problematic or incomplete about this.

So we must give up thinking about the nature of things and observe their relations in mathematical terms—as if we could possibly describe the relations between things without first understanding in words something about the things themselves! Yet the fact is that we cannot even *see* a thing, let alone determine its relations, without taking it to be a certain *kind* of thing possessed of its own characteristic qualities (Brady 2002). The question is only whether we will accept uncritically our half-conscious assumptions about the substantial nature of things—as when, for example, we imagine subatomic particles to be very tiny bits of the qualitatively familiar stuff we deal with every day (an imagination that has caused no end of grief to physicists)—or whether we will instead raise these notions to full consciousness, where we can subject them to proper criticism.

To one degree or another, our science always does have real content, and whatever their philosophical disclaimers, scientists always do believe they have learned something about what Dijksterhuis dismisses as "the true nature of things." And when we lay down our measuring instruments and let go of our high abstractions long enough to examine critically this meaningful content of our theorizing—when we try to understand the entities or processes, the phenomena, without which our mathematical formulations give us no knowledge of the world—then we find ourselves facing three closely interwoven aspects of the world as it becomes known to us: it is irreducibly qualitative; it is a manifestation of consciousness; and it is thoroughly contextual.

Qualities

Just as we all understand words as bearers of meaning in the practical sense that we successfully read the meaning of texts, so, too, we all understand qualities. In its perceptual immediacy, the world is nothing but its qualities; this perceptual content—as opposed to the pictures we have formed of our own theoretical abstractions—is what I mean by "qualities."

Try sitting outdoors in a natural landscape for half an hour. After quieting yourself and becoming as receptive as possible, ask yourself—not regarding your own thoughts, but regarding the content presenting itself from the world: What is the character of this content? It may present you with opportunities to measure and derive quantities, but what are you immediately given? What is it that, only subsequently, you are free to measure? From the sky and the distant hill to the grass, pine needles, or soil beneath your feet, is there anything here at all of which you can say that it *is* measure or can be fully expressed in terms of measure? (How many of us, during years or decades of creative work, will put the question, "What do I meet in the world?" *directly to the world,* as opposed to thinking *about* the question in our studies or laboratories, with our thought mediated and perhaps falsified by a vast network of mental abstractions?)

Or try subtracting from the content of your observation everything except measure. In the case of the tree over there, remove the green of the foliage, the gray of the bark, the smell of sap, the rustling of leaves in the breeze, the felt hardness of the trunk . . . and what do you have left? Nothing at all—certainly nothing to measure. You do not even have geometric form, since without light and color there is no visible form, and without the different qualities of touch there is no felt form. Measurable form is not something independent that we proceed to flesh out with qualities; it subsists in nothing but perceptual qualities themselves.

You may want to say that the quantities we abstract from our qualitative experience of the world point us toward a reality hidden behind the world we perceive. But unless you can say something about this hidden reality—unless you can characterize it, giving your quantitative constructions some sort of content—where is your science? And how will you characterize this content without appealing to qualities?

Whereas (in Russell's and Feynman's terms) mathematics and logic as such give us nothing to talk *about,* it appears that the content we do have to talk about—the content scientists in fact talk about all the time—consists of nothing other than the world's qualities. So, despite our entrenched habit of refusal throughout these past several hundred years, shouldn't we undertake an explicit inquiry into how our work with qualities can be as truthful and meaningful as possible? Perhaps these qualities are the world's native way of presenting itself—not a terribly strange hypothesis, given that we cannot imagine any other manner of presentation. In any case, we could not reasonably deny the hypothesis except by investigating the qualities in their own terms.

These terms are not particularly obscure; they simply refuse to conform to our preferred scientific stance. An elementary quality such as "red" proves maddeningly elusive when our aim is to pin it down. My red shirt turns out to be a different color depending on the lighting and on the other colors around it, as well as on the state of my own eyes. Similarly with the qualitative nature of an entire complex organism: we recognize a *single* species-nature in a lowland spruce tree and an alpine one, but this common nature comes to dramatically *different* expression in the two cases. So qualities exhibit the one feature the logician must not tolerate: no quality is "just what it is and not something else." Qualities interpenetrate one another, manifesting themselves differently in every different context.

Since qualities lack the sharp-edged, yes-or-no, unambiguous character of logic, the question to ask about any qualitative description of a phenomenon is not so much the simplistic "Is this precisely true (yes or no)?" as the more challenging "How fully and in what way does it reveal what speaks in the phenomenon?" Does the sloth's apparent listlessness indicate something fundamental about its character, and if so, how does this relate to its clinging tendency and its openness to its environment? (See chapter 12.) Serious qualitative descriptions are never merely true or false; rather, they exhibit more or less expressive depth. They give us a more or less satisfying, a more or less penetrating, insight into (and feel for) what a phenomenon is *like.* When we are reckoning with qualities, questions of similarity are more central than questions of identity. It's one thing to record the contours of someone's face as a set of precise spatial coordinates, and quite another to notice the distinctive *character* of the face. Often, however, we can read very little of this character in a frozen snapshot. That's because qualities are dynamic, not static. What they *are* is their inner movement, their manner of exchange and mutual interaction, so that we can catch them only in flight—by moving in a like manner along with them. They leave behind every effort to grasp them and pin them down. A sculptor of stone succeeds only by suggesting movement. Even in depicting a massive rock *as a rock* we must somehow capture a movement of profound rest, an ageless silence that is itself speech.

We can certainly learn to know qualities. However, our inner activity in taking into ourselves a particular quality involves much more than the play of abstractions over the surface convolutions of our brains. We can move with qualities only by acquiring some of the artist's sensitivity, whereby our mobile feelings and our active will are playfully engaged. We experience qualities with our whole being, discovering, for example, that this color has something cool about it, that one something aggressive, and the other one something calming—characteristics of the sort that great artists have always had an objective ability to work with.

To look at the world with an openness to its qualities is to ask, "What kind of phenomenon meets me here? What is it expressing through the distinctive way it summons and coordinates the world's lawful grammar? What melody of its own is this phenomenon picking out upon the mathematically tuned world-lyre?"

"Quality" is in fact an approximate synonym for "meaning." But we usually speak of qualities when we are referring to the world, and we speak of meaning when we are referring to language and thought. The two usages are closely intertwined. The way we reduce the world to atomic *things without qualities* is by reducing our descriptive language to the atomic terms of *logic without meaning.* That is, we can obscure the qualitative character of the world only by obscuring the meaningful character of our words. But we never fully succeed in this. The world remains word-like because it is full of the meanings of language, just as our words remain world-like because they are full of the qualities of the world.

Consciousness

Everything we've noted about qualities points to the fact that they are expressions of consciousness. This is hardly controversial; the reason why the scientist fled qualities from the very beginning is that they "reside only in the consciousness" (Galilei 1957, 274). But given that the only "place" we have for experiencing and knowing the world is in consciousness, and given that we evidently do gain at least some real understanding of the world, the obvious thought to entertain exactly parallels the one we entertained a moment ago about qualities: perhaps the consciousness through which alone we can come to understand is in fact well suited to understanding, and for a very good reason: in the world our consciousness meets something like its own activity, something akin to its own nature. With our wide-ranging potential for conscious experience, we are ourselves expressions of the cosmos. Is it surprising, then, that we should be able to give conscious expression to what speaks in the world?

In some scientific quarters a thought like this proves outrageous, while at the same time some of those most envied of scientists, the physicists, speak casually of consciousness as in one way or another fundamental to the cosmos. Even at a time much less hospitable to this thought Sir Arthur Eddington, alluding to the problem we have been considering, could write: "[Our knowledge of physics] is only an empty shell—a form of symbols. It is a knowledge of structural form, and not knowledge of content. All through the physical world runs that unknown content, which must surely be the stuff of our consciousness" (1920, 200).

Presumably he means "unknown" only in terms of the accepted, one-sidedly quantitative ideals of science, for if there is one content we can know at least to some degree, surely it is the content of our consciousness.*

---

* The resistance to Eddington's conclusion is even odder when you consider how many authorities in different fields loudly disavow the Cartesian diremption of matter from mind. The situation becomes more understandable only when we realize how thoroughly Cartesian these authorities remain: they take their stand firmly astride the fractured Cartesian bedrock, accepting the division of things in Descartes' terms and then hoping only to make one side of the problematic divide disappear by reducing it to the terms of the other. A real solution will be found only when we go back and refuse the split altogether, finding another way forward. And this way will include the recognition that the world has a word-like character. Only in language do we find the marriage of inner and outer in a way that overcomes all the conundrums of the mind-body dichotomy. But appreciating this solution can require agonizingly hard work when you have been raised, as nearly everyone in our culture has, upon Cartesian habits of thought. My own path away from these habits was blazed by the philologist and historian of meaning Owen Barfield. See, for example, the works by Barfield listed in the References.

Can know, I say. If we choose to continue ignoring this qualitative content—if we refuse to become familiar with it and to begin learning how to approach it in a more or less disciplined and systematic manner—then it will certainly remain Eddington's "unknown content."

The prevailing refusal on this score can hardly be disputed. When as scientists we analyze sound—whether of a volcano or a musical performance—solely in terms of pressure waves and other mechanically conceived processes, our terms are, curiously enough, as fully available to a deaf person as to someone with good hearing (a point somewhere made by the German physicist and educator Martin Wagenschein). In fact, the ideal of rigor within the hard sciences generally aims, rather impossibly, for the use of terms understandable by someone who has no conscious perception of the world whatsoever. But such a person, if he actually existed, would have no world in need of understanding! If, on the other hand, we do have a world to understand, it is a world whose nature is to present itself within consciousness.

Context

We gain a kind of unqualified crystalline clarity by filtering our perceptions of the world through a web of logically precise abstractions. Even space and time become, through analysis, a collection of discrete points or discrete instants. But the quantifiable crystalline clarity we thereby achieve belongs to our perceptual filter and not to the world. Dazzled by this clarity and fixity, we become blind to context.

Go out again into a natural setting, sit down, and spend a while taking in everything you can see, hear, feel, and smell. Then ask yourself: does this world, in any meaningful sense, consist of discrete points or instants of time?

You will be hard put to find any justification in observation for these abstract notions. The world and its events present themselves—stunningly, when you compare your actual experience to the various theoretical ways of thinking about the world—as one seamless whole. Points and instants flow into each other, participate in each other, and cannot be clearly separated from each other. Likewise, the seemingly incommensurable "data" of our sight and hearing, our smell and touch, yield moment by moment a single, unified image of the world. Pick any visible object—a tree, say—and try to isolate it cleanly and without ambiguity from everything around it. It cannot be done.

Again, this is hardly controversial. The entire discipline of ecology was founded upon the awareness that organisms are an expression of their environment, and the environment is an expression of its organisms. At the largest scale, the earth's atmosphere, as it once existed, would have been poisonous to today's living forms, and only through the influence of the evolving life forms themselves has it become what it now is (Lovelock 1987). The earth's breathing and that of its creatures is one breathing, and the organism meets itself in its environment.

But it is not only organisms that require a contextual understanding. Every entity, process, and law described by physics gains its real content only by means of its context. We may be able to discern *in* a process a certain quantitative lawfulness that is invariant from one context to another—because the quantities have had all context and phenomenal content stripped from them. But while this absolute sort of lawfulness may be abstractable from the physical process, the observable content itself is never invariant or subject to necessity in the way we take our universal laws to be. The reason we have to abstract the content away in order to arrive at the mathematical "explanations" is simply that the content of real events is not explained by the mathematics alone. In a beautiful meditation, physicist Georg Maier offers examples, some of them very simple, of the fact that the world's material processes can be understood only contextually (Maier et al. 2006, chapter 10):

- "Warm air rises"—and so it does in a closed room, where you will find the air warmer near the ceiling than at the floor. But in the open atmosphere the air usually gets colder with height. You can understand the difference only by considering the two different contexts, one of which limits the upward movement of air, while the other does not.
- Gravity has very different effects, and must be described in different ways, depending on whether you are walking on the solid earth, "floating" in orbit around the earth, or swimming in a lake. The different effects extend even to the question of whether your bones will be subject to a dangerous loss of mass—something of concern to long-term inhabitants of orbiting space stations.
- If you place a lighted candle inside a jar and then accelerate the jar (along with its atmosphere and candle), you will find the flame leaning *forward* in the direction of acceleration, a behavior "contradicting" our more common experience with accelerated objects.

An Important Distinction

These examples will seem either trivial or profound, depending on our ability to discern the subtle distinction they require. The statement "warm air rises" refers to observable behavior in the world, and therefore, construed as a universally valid law, it fails. All you need to do is to change the context, and a different behavior results. This is true of any law presuming to specify, in unqualified terms, what real things will actually do. True understanding must be thoroughly contextual, so that in different contexts the phenomena will bring their lawfulness to different expression. We overcome this contextuality only by retreating from a description of phenomena to a statement of "grammatical" regularity abstracted from all concrete and particular reference.

You can see the retreat in a law such as Newton's universal law of gravitation, which might be stated this way: "Every particle of matter in the universe attracts every other particle with a force directly proportional to the product of the masses of the particles and inversely proportional to the square of the distance between them."

Here there is no longer an assertion about what any particular things will actually do ("warm air rises"), and this is why the law escapes falsification by different contexts. It doesn't tell us about contexts; it is a decontextualized statement. Its potentially meaningful terms—for example, "particle" and "force"—remain more or less blank; they are the seemingly inexplicable mysteries of today's physics, leaving us only with the certainties of the mathematics. The "attraction" it speaks of is not a specific, observable behavior of any sort—not, for example, a movement of objects toward each other—but a grammar that any actual movements will be found to respect. Real bodies moving according to this grammar may approach each other, spiral around each other, or move directly away from each other.

The actual behavior of things in the world is always an expression of context. What Maier says of a gas can be said of everything we encounter in nature: it "is so intimately entangled with its environment that its phenomena can be accounted for only as part and parcel of a greater whole." If we want a lawfulness bearing on such contexts, then we will have to look for—what else?—a contextualized sort of lawfulness. The coherence to expect is more like the coherence of a picture or image than of isolated entities. It is more like the coherence of a language context (which plays into and colors every word) than the coherence of a logical or mathematical proposition. And therefore the unity of the context, or whole, is a conceptual prerequisite for our understanding of the part.

Qualities already imply context. This is because they refuse to be "just what they are and not something else," but instead interpenetrate and share something of their identity with each other. Discrete, qualitatively featureless particles can exist only in nameless, side-by-side aggregation; they can never give us the kind of contextual unity that plays into, modifies, and binds together the various elements of the context. In order for there to be a true context, something qualitative must reach across and penetrate all the elements, shaping each of them to the character of the whole. In physics, it may help to think of the hologram. In the life sciences we discover a similar but more living unity of the organism, which one of us (Craig Holdrege) alludes to in his study of the sloth when he writes that "every detail speaks 'sloth'" (chapter 12). And, in fact, it makes sense to say that contextual wholes are *spoken,* because they are a weaving together of qualities, or meanings.

Holograms and organisms, images and contexts—these are a long way from the kind of lawfulness we see in Newton's universal law of gravitation. However, nothing requires us to give up our extremely useful inquiry into nature's formal grammar. But the many conundrums into which this inquiry leads us—conundrums widely recognized whenever scientists temporarily extricate themselves from their dense mesh of theoretical abstractions, face nature herself, and try to understand what they have been talking about—will remain insuperable obstacles to progress until we can begin to explore what a contextual and more imaginal form of understanding might require of us.

Reified Equations

Here we need to pause and consider the instinctive objection that almost inevitably arises at this point, rooted in stubbornly entrenched habits of thought: "It may be true that a universal law such as the law of gravitation cannot tell us everything that will happen, say, among the objects of the solar system. But that's because there are other laws at work as well. If you add in all these other laws, then, at least in principle, you will understand everything that happens. What can you point to that escapes this all-encompassing lawfulness?"

The short answer is a simple reminder: I have not been suggesting that anything needs to violate or escape the universal laws of physics—no more than a meaningful sentence needs to violate the rules of grammar. But the necessities of a sentence's grammatical form are insufficient to determine what the sentence is saying—they do not give us its content—and neither do the formal necessities of mathematically stated laws give us the content of the world. Where we have such content, it speaks forth its own coherent meaning, and while this meaning may always respect an underlying formal grammar, it can never be reduced to such a grammar.

But this short answer requires expansion. Think of the movements of the heavens, which perhaps are what most naturally come to mind when we imagine the determinism of physical law. Perfectly timed eclipses, precisely targeted space probes, the regular rhythms of day, month, and year—certainly these are real phenomena, and yet we commonly manage to predict them with extraordinary accuracy. Could any phenomena be more fully determined by mathematical law than these?

Well, again, the point is not that mathematical laws must be violated. Nor is it that there must be some element of randomness or wild, lawless disorder in the cosmos. Contextual coherence, after all, is not randomness and disorder. But neither can its significance be expressed in purely quantitative or formal terms. It is a different—a *meaningful*—kind of order.

How easily we overlook the ways we empty the world of its content! When we think of the heavens as explained by the mathematics of the universal law of gravitation (and other laws), we have reduced our explanatory expectations almost to nothing. Sun, planets, and moons are transformed into anonymous point-masses whose "character" or "behavior" consists of nothing but their trajectories through space. These bodies have become in our minds little more than blank reifications of their governing equations. The phenomena themselves have almost completely dropped from view. It's as if on earth we looked around at people, muskrats, squash plants, clouds, boulders, and springs, and saw no diverse challenges to our understanding, but only the one task of calculating the diverse spatial trajectories of objects.

You need only reflect upon all the scientific disciplines arising from our experience on earth to realize that when we think of the moon or sun as mere points in motion, we have blocked from our view virtually all the reality of these bodies. If you ignore everything except points in motion—everything constituting the expressive reality of a phenomenon—then it is not the phenomenon you are describing. You are simply using an image (say, of the moon) as a token to stand in your imagination for the lawful grammar that, as we have already recognized, can be abstracted from the physical world. You may choose not to interest yourself in anything more than this grammar, but such a self-imposed restriction hardly constitutes evidence that grammatical knowledge is the only knowledge available to us.

Earlier generations spoke of various *influences* streaming in from the heavens, and of the humanly relevant *dispositions* of celestial bodies (or beings), and of the *lunatic* or *mercurial* nature of people or events, and of the heavens declaring the *glory* of God. If, after peeling away the layers of superstition accreted around such notions, we are to assess whatever validity they may have had—or even if (which must come first) we are to understand what sort of thing the ancients might have meant by such terms—we will have to look beyond a mere grammar of movement and open ourselves to the world's qualities. The modern denial of these qualitative judgments is not science; it is a simple expression of the choice within science to have nothing to do with qualities.

What Is a Force?

Even if we start with the words we commonly use in stating our most rigorously quantitative physical laws, and if we take these words as really meaning something, we are immediately carried toward a richly qualitative world. At mid-twentieth century the philosopher Kurt Riezler, speaking about the concept of force, chided physicists with these words: "You use the word force and, when queried, you define it by law, field, and vector; but what you really have in mind is the force you feel in commanding your muscles" (Riezler 1940, 11–12).

Can we gain an adequate scientific understanding of gravity except by referring to the willful use of our muscles or our experience of pressure? A little reflection will convince us that the answer is no. Of course, many scientists will react initially to the question by citing the purely objective relationships of moving masses—relationships given in strictly mathematical terms. And it's true that objects changing their positions in space may give us certain mathematically describable relationships. But so, too, can points on a piece of graph paper. No one takes these points to be exerting a physical force upon each other. Neither could we think of planets as exerting a force *unless we had independent concepts of mass and force.* As the graph paper illustrates, the mathematical relationships alone do not give us such concepts. Think about it all you wish, but a force is something real in the world, and you will never find a concept for it except through your own experience of the world's forcefulness.

This experience, like all our experience, occurs within consciousness. And it is an indication of the radical and unexplored possibilities of a qualitative science that we cannot say a priori that our conscious experience of bodily forces is unrelated to our experience of the "force" of a personality, the "force" of suggestion, or even that attractive power, or "force," of love that some of the ancients imagined to be at work in the descent of heavy bodies toward the earth. Such possibilities may be crazy or not, but confirming or refuting them would require a kind of devotion to our experience of the world that we long ago lost interest in sustaining.

In any case, what we cannot escape is that without *some* sort of experience of force within the inner domain of our own consciousness, we have no meaning for the scientific concept of force. Of course, the law of gravity is not meaningless, and we heard Riezler explain why: we can't help bringing our conscious experience to the law, even if this experience remains more or less unacknowledged and therefore is never subjected to proper scientific criticism.

Logic and Scientific Explanation

The common notion of cause and effect in science is a denial of the experiential basis of our understanding. Reducing the expressive content of a phenomenon as far as possible to a few isolated elements possessed of perfectly definable relations, we imagine a "closed system" immune to outside influences. We shift from imaginal thinking to abstraction, from recognition of qualitative experience and the mutual interpenetration of elements to the search for isolated, well-defined parts. Then we say, "This makes that happen," and we try to get a meaningful larger picture by summing these cut-off, individual instances of what we imagine to be cause and effect. In our quest for certainty we want our "things" to play the role of reified symbols in formalisms of mathematical logic, and therefore we must empty the things of their content and isolate them from their context.

I have already noted the fundamental error in this: formalisms can give us a grammar for our description of events in the world, but the grammar by itself is helpless to provide the description. It can neither generate nor explain whatever descriptive content might be expressed in accordance with the grammar. If we discover an underlying grammar in the world's phenomena, then our task as scientists is to apprehend in full consciousness the speech from which the grammar has been abstracted. In this speech the meaning of a contextual whole takes qualitative hold of the part and raises it to a participation in the whole. And so, rather than emptying our terms of experience and meaning, we need to fill them as fully as possible, taking care to see that the experience is truly an experience *of the world* and not merely of our own unconscious impulses.

Paradoxical as it may sound, to the degree our convictions carry a sense of logical certainty, they are built on illusion. Or even more paradoxically: the only thing a sense of logical certainty in science *proves* is that we have disconnected ourselves from scientific understanding. To mistake certainty about the internal validity of a logical structure for certainty about the real-world content we project onto the structure is to have lost sight of that content in its own terms. We allow the logic to tyrannize over us, as when we imagine the "law of gravity" to dictate the actual expressive behavior of things, rather than merely to indicate a grammatical form that helps to make expressive behavior possible.

Because logic never does demand or explain a particular expressive content, the fact that it seems to us to do so bears witness to a sense of explanation on our part that is unrooted in the world—unrooted in the actual presentation of things. If we believe the things we are talking about are determined by their grammar, then the real determination of our thoughts is escaping our notice. This points to the ultimate paradox, which is that an improper trust in the logical force of our scientific explanations puts us at the mercy of irrational powers.

If you doubt that perfectly reasonable people holding perfectly reasonable opinions can be subject to irrational powers, a bit of self-reflection may prove helpful. Why is it that virtually all of us, however highly educated and critically minded, continue to hold many or most of the beliefs and values characteristic of the immediate ethnic, religious, educational, ideological, and cultural contexts in which we were raised? We may engage in sophisticated critical analysis and assessment of our beliefs and values, yet the outcome of our argumentative logic clearly depends in substantial part upon which groups we are members of. The basic and largely unexamined meanings we have inherited provide a perfectly compelling logic for our intellects to work with. Enamored by the force of this logic, we fail, in our critical reflection, to reach down to the unconscious meanings in a way that might lead us to very different understandings. Logic itself, stripped as it is of the meanings from which we abstract it, can never lead us to a fundamental reevaluation of those meanings.

I do not say this in order to play the skeptic regarding all belief and value, or to argue that we never have any justification for holding convictions characteristic of our community. I am only pointing out that we do well to realize how naturally our routine use of logic, whether in arguments with a spouse or in defense of hallowed dogma, can be governed irrationally and from below. And this is at least as true of the scientist's materialistic dogma as it is of any other religious dogma. A satisfying sense of logical compulsion always puts the truth in peril.

To Explain or Portray?

If the kind of logical or grammatical necessity associated with precise scientific law cannot give us an explanation or any full understanding of the world's phenomena, where do we find understanding? We need first of all to accept that an understanding of whatever content is conveyed by a grammatical structure is very different in character from an understanding of the grammatical structure itself. It's the difference between a flat and a many-dimensioned view of things. In particular, we are brought to a different notion of causation as soon as we try to rise above a merely grammatical sort of understanding.

In aesthetics and in the notion of "formal causation" tracing back to Aristotle, the *formal cause* of a phenomenon or work of art is its unifying shape or form. But this shape is not taken to be a mere distribution of mathematical points within a spatial grid; rather, it is the overall expressive gesture of the thing. This older conception of cause points us toward the qualitative form or meaningful patterns, the governing unity, according to which phenomena unfold rather as a spoken sentence progressively unfolds to express an antecedent governing idea—an idea that informs and transforms the individual words, shaping them to itself. This meaning of "formal" is nearly opposite to the "formal" and "formalism" I have been employing until now.*

The more usual way of thinking about scientific explanation ("This makes that happen") can be useful as long as we realize we're dealing with approximations and that the more we approach an absolute precision and necessity in our cause and effect, the more we have abandoned the context with its expressive character, so that, in the end, nothing of any particular phenomenon remains. We redeem the approximations by realizing that they *are* approximations and by allowing them to clarify details which we then enliven by bringing them back into qualitative connection with the meaningful whole.

This is very much the way neurologist Kurt Goldstein approached the "mechanical reflex" in his several-decades-old and important book, *The Organism* (more recently reprinted with a foreword by Oliver Sacks). Goldstein looked at the various ways we analyze organisms into rule-based, mechanical parts and then try to reconstruct the whole from these parts. It never works. He assesses the reflex in humans and animals, showing in exhaustive detail that the "simple-minded" reflex mechanisms we so easily imagine don't really exist. For example, slight changes in the intensity of a stimulus can often reverse a reflex; a reflex in one part of a body can be altered by the position of other parts; an organism's exposure to certain chemicals, such as strychnine, can reverse a reflex; other chemicals can completely change the nature of a reflex; fatigue can have the same effect; consciously trying to repress a reflex can accentuate it (try it with your "knee-jerk" reflex); and so on without end.

Goldstein showed that the reflex is an artifact of our own stance as researchers, whereby we conceptually and experimentally isolate one part of an organism, cutting the part off from its whole. Moreover, he finds that higher organisms, including human beings, are much *more*

---

* There is a common misunderstanding of formalism and of the relation between form and content. Many believe that formalism concerns itself with the form rather than the content of things or events. But the fact of the matter is that formalism, as the one-sided drive I have been discussing, abandons both form and content. There is no form without content; form can only be the form *of something*. When the content is left behind, so is any form it might have displayed. Because our thought is taking ever more precise form, we may think we are laying hold of the form of things ever more precisely. But, as I have tried to show, what we are laying hold of is less and less of anything at all. Because there is progressively less content, there is also less and less form. Form and content are distinguishable but inseparable aspects of whatever is there.

likely to show approximations of reflexes, because it is we who can allow parts of ourselves to become isolated and de-centered. (That's what many procedures of medical assessment are all about.)

> Human beings are able, by assuming a special attitude, to surrender single parts of their organism to the environment for isolated reaction. Usually, this is the condition under which we examine a patient's "reflexes." . . . But [regarding the pupillary reflex] it certainly is not true that the same light intensity will produce the same contraction when it affects the organ in isolation (as in the reflex examination) and when it acts on the eye of the person who deliberately regards an object. . . . One only needs to contrast the pupillary reaction of a man looking interestedly at a brightly illuminated object with the reaction of an eye that has been exposed "in isolation" to the same light intensity. The difference in pupillary reaction is immediately manifest. (Goldstein 1995, 144)

In sum, we arrive at the law of the reflex only by isolating a separate part of the organism and confining our attention to this part in disregard of the whole. The only way we can achieve such isolation is by draining the context of its interpenetrating qualities, such as the quality of *interest* and the corresponding qualities of the eye in Goldstein's example of the pupillary reflex. The fullness of reality fades away, leaving the kind of logical skeleton we so easily conceive as a mechanism, with its separate, well-defined parts.

This is to suggest that we can best understand exact, fully determining causes as the skeletal ghosts of formal causes. They are more or less ineffective when juxtaposed with the sinews and countenance of reality, and reveal their impotence when we try to make them account for this reality. Johann Wolfgang von Goethe was pointing to this inadequacy of cause-and-effect explanation when he remarked of his pioneering morphological research that "its intention is to portray rather than explain" (Goethe 1995, 57). Goethe's idea seems to be that description—or at least description of the right sort—itself constitutes understanding. This is implied more strongly in another of Goethe's oft-repeated koans, which anticipates a great deal of modern thought: "Everything in the realm of fact is already theory. . . . Let us not seek for something behind the phenomena—they themselves are the theory" (307).

It's obvious enough that we cannot describe anything well without having a good understanding of it, and this understanding informs the description. Goethe's sage advice sounds anemic only when we cannot let go of the misplaced hope that the world might be grasped and explained in the way a logical structure, once we have purged it of descriptive content, can be grasped and explained. This is to forget that logic helps us on our way toward understanding only when, in the very act of performing its clarifying function, it sacrifices itself to the expressive content from which we drew it out. This sacrifice is the reverse of that crucifixion of the world-text upon a scaffold of logic to which I referred earlier.

To prefer portrayal over explanation is to reject the *one-sided* (and never fully achievable) drive to isolate restricted contexts and precisely definable causes or laws. It is to refuse to lose sight of the interpenetration and mutual participation of things, even while accepting the necessity for narrowly focused excursions. It is to let go of explanation as something *fixed,* as something we can *have,* which easily becomes a dead weight upon further inquiry. A portrayal requires a stronger, more full-bodied inner activity on our part in order to hold everything together and grasp its coherence; the portraying is something we must *do,* not only with thought, but also with feeling and will. We have to trace the fluid, complex way in which one contextual *picture* metamorphoses into another instead of the vanishingly simplistic way in which isolated (and therefore impossible) things univocally affect other isolated and impossible things.

We have to engage in this inner activity because it is the only way we can move harmoniously with the activity we encounter in the world. It is the only way we can truly understand the qualitative language of nature, which is at the same time the meaningful language of our own being.

• III •

The Wholeness of the Instrument

We are creatures of the word, inhabiting a world that can be understood only as speech or text—even if we prefer to notice only its blank, unspeaking grammar (which, nevertheless, presupposes the speech). Our own communication depends upon the word-like character of the world; if we did not find word-stuff all around us, we would have no material for our own words. Nature presents us, not with blank, mute, disconnected objects, but with expressive *images*, and such images are the native elements of story, song, and poetry. Even at the level of "mere" sound we can say: only because every sound has its own gestural and significant form—only because it speaks with its own qualities—can we recognize it as a distinctive element and employ it for our own speech.

Further, even where we have reduced speech to the high abstraction of alphabetic text and have rendered the connection between word-signs and their meanings perfectly arbitrary, a speaking quality of the signs themselves is prerequisite to our apprehension of them as words. If we did not perceive and feel the different qualities of a horizontal and vertical stroke, and again of a vertical and circular stroke, we would be unable to distinguish one alphabetic character from another.

This last point, humble as it may be, sums up everything I've been saying in this chapter. It is also extremely difficult for people of our day to grasp, and is worth a great deal of reflection.

Gestures of the Morning

If we were given a set of mathematical coordinates defining the pixels, or points, of a pen stroke, these coordinates would remain meaningless; we could not recognize them as any sort of unity or as any specific thing—not until, with the aid of our knowledge of coordinate systems, we pictured the points as constituting a significant form. What is hard to appreciate is that we cannot recognize anything except by recognizing it as a particular *kind* of thing, having some sort of apprehensible character. There is always an aesthetic judgment at work. We can distinguish a horizontal from a vertical line only because, in an objective and cognitive sense, we can *feel* the difference between them—we can experience the difference rather in the way we experience the difference between our own arm held first horizontal and then vertical, and in the way both actor and audience experience the dramatic contrast between an outstretched and an upraised arm; in every specific context they speak differently. The two gestures have different expressive potentials (Rozentuller and Talbott 2005).

The natural world is nothing but such gesturing, even if at a vastly more profound level than our own gesturing—a level where the gesturing is at the same time a power of physical manifestation. We can know that something is there only so far as we find ourselves "gestured at" in a recognizable way. It may be worthwhile to take a brief look at one or two of the countless "alphabetic strokes" of nature with which any qualitative science is likely to have to reckon. I will not speak here of any body of science, but merely of possible elements belonging to a language of scientific description.

Not so long ago, soon after awakening on a cool morning, I stepped out of my darkened home into the radiant and golden warmth of the newly risen sun. Having had nothing particular in mind, I was suddenly and unexpectedly moved by a feeling I can describe (inadequately) only as one of expansion—as if, in a spirit of rejoicing, my arms and my entire being were opening outward to embrace the fullness of the world. Not being one to live at all vividly in his perceptions, I was struck by the force of the sensation, and began to wonder whether it spoke in any objective way about the morning and sunrise. Was I experiencing a significant element in the language of nature, or just the incidental noise of my own body and psyche? Does nature speak forth the dawn of a new day in a unified language of aesthetic gesture? Is there, for example, any possible justification for the common sentiment that nature in some way *rejoices* in the dawn?

Over the following weeks I focused on a few simple gestures—for example, the expansive opening to the world I had experienced, and the ascending movement we see in the rising sun. And it proved useful to contrast these with more or less opposite movements.

However much I might have been inclined to dismiss my own sense of enlargement on meeting the sunrise, I could hardly dismiss the fact that *nature* was going through something analogous. In a literal, physical sense, nearly all substances—rocks, lakes, plants, the earth itself—expand under the heating effects of the sun, and they contract again as the environment cools at sundown. The atmosphere, too, dilates under the sun's influence, which also sets in motion rising currents of air. We often see a morning mist rising from sun-lit ponds. And if there is a cloud type most characteristic of the day, surely it is the expanding, upward-billowing cumulus cloud. At night the cooling atmosphere "settles down," and we may find a layer of fog pressing against the earth. Cyclonic storm systems, which tend to contract and lose intensity at night, begin to grow again in geographic extent in the morning.

Living things exhibit similar gestures. Perhaps most noticeably, many flowers and leaves open outward in the morning and close in upon themselves at night. In herbs and trees, the morning sun draws the sap upward; water, too, ascends from the roots, engorges the leaves, and then evaporates outward and upward into the atmosphere. We ourselves greet the day by stretching ourselves: our chests swell, and we extend our arms toward the periphery as we prepare to meet the world again. At the moment when we awaken and look out upon a sun-lit world, it is easy to experience how our psyche is transformed, gaining a certain outward-oriented solidity and spaciousness as we are enlarged by our surroundings.

And the sun itself? It *radiates.* Here I am not referring to the falsely imagined "rays" supposedly traced by particles of light. I am speaking of the *gesture* of light—a gesture we can perceive directly. At dusk, let your eyes rest for a while upon the darkness of a valley and wooded hillside, then raise your gaze to a remaining patch of light sky. Once you have learned to still your thoughts *about* what you are seeing and instead to use the perceptive capacities of your entire organism, you can experience the radiating gesture of this light—sometimes almost explosively. Light that is *too* bright strikes us so forcefully as to make us recoil and shield ourselves from injury.

I remember hearing people talk, earlier in my life, about the two different expressive characters of sunrise and sunset. At the time this struck me as misguided: "The two occurrences are exactly symmetrical, with the sun shining from one horizon or the other through the thickness of the atmosphere. The difference between east and west can hardly be crucial. You can have beautiful red sunrises just as you can have beautiful red sunsets. Where is any essential difference between them?"

This was unutterably foolish of me. I was thinking in terms of static snapshots, entirely forgetting their context. In reality the two events are polar opposites. Leaving aside the fact that the constitution of the morning atmosphere tends to differ considerably from that of the evening atmosphere (for example, there is typically more haze or dust in the evening), there remain the most obvious features: the morning sun is rising while the evening sun is setting; the day is brightening and warming, or else it is darkening and cooling.

Try a simple exercise. Stand upright with your arms at your side. With your consciousness as quiet as possible, attending only to the inner qualities of your movement, very slowly swing your arms in front of you, palms upward, as if you were following the rising sun with your hands. Then pause, turn your palms downward, and let your arms slowly perform the reverse movement, descending to the starting position. (You will be forgiven if you find something almost reverential in the exercise.) With any attention at all, it is not hard to experience the very different character of these two motions. With the one we easily feel (among many other things we might pay attention to) a sense of anticipation, of beginning, of active engagement, perhaps even of celebration. With the other there is a calming, a coming to rest, a sense of completion—and again, it may be, celebration, only now it is celebration of fulfillment more than expectation. At the very least, we have to say that a sunrise and sunset are as different as these two expressive movements.

It's important to avoid a kind of wooden oversimplification, as if we were dealing with fixed elements of logic rather than living gestures of the world. Every slightest alteration in the overall constellation of a gesture makes for a *different* gesture. In fact, it's probably impossible ever to perform exactly the same gesture twice. If, instead of asking you to let your arms descend directly in front of you as you followed the movement of a sunset, I had instead suggested that you position your arms a little more widely apart, perhaps bending them more at the elbow, then the restful and calming aspect of the gesture would have been accentuated. If I had told you to relax your arms completely, letting them fall limply under the influence of gravity, the descent would have spoken more of heaviness than of rest. And if I had asked you to hold your arms rigidly straight during the downward movement, with elbows locked, the sense of rest would again largely have been lost.

The failure to recognize the multivalent potentials of every abstractly understood movement ("descent")—and also of every other kind of gesture, such as that of a color or a sound—has resulted in a great deal of nonsense being written about the lack of any universal or objective language of qualities. Every gesture is concrete and contextualized, and therefore unique, but contextualization is not the same thing as arbitrariness. Anyone who has worked with gestures—the sculptor, for example, and painter—knows that she is working with a language not only of boundless expressive potential but also of great definiteness and consistency. While qualities are fluid and interpenetrating, modifying one another and lacking any fixed and static identity, they nevertheless have a vivid character that we can enter into and work with.

Returning, then, to our theme: clearly, whatever coherent "morning conversation" may take place among stone, flower, cloud, and the rising, radiating sun, the language of this conversation is not in any primary sense the language of universal physical laws, in the usual mechanistic and mathematical sense. Whatever expressive unity may exist between the expansion of a stone and the opening of a flower, we cannot portray it merely by citing particular mechanical principles they happen to share. Yet the question remains: while speaking in the distinctive language of their particular substances and organization, are they contributing their own harmonies to an integral symphony of the morning we can recognize throughout nature?

Of course, if we attend only to mechanical principles, then the various gestures I have cited can never even occur to us as a challenge for our understanding. However, as I have tried to point out, all our science is grounded in one way or another in our experience of just such gestures. Even the world's basic materiality, grounding the scientific concept of mass, gains content for us only by virtue of a certain inner movement expressing something like denseness or compaction, and also resistance. Without this inner experience, we would have no content for the concept. It wouldn't mean anything; we would not know how to go about exploring its mathematical grammar because we would have no "it" to guide us in picking out what to measure.

Once we have recognized the real content of our scientific language, we can hardly turn away from the kind of question I am raising here, however unfamiliar it may seem. The question is not, "Do the stone and flower make expansive gestures under the influence of the sun?" This is simply given and cannot be doubted. Rather, what we want to know is to what degree the collection of gestures I have cited can, when united with a great deal more understanding of morning and evening contexts, speak in a coherent, aesthetically unified way.

Don't think this is a simple or obvious matter. I may have selected my examples very clumsily, just as my report of a conversation with a friend might miss the central point by collating some more or less incidental remarks based on an external or merely grammatical resemblance. And, in fact, my references to opening, radiating, expanding, and ascending movements have been rather abstract and general, without much reference to the concrete and varying qualities of the movements, and without extension to the broad range of phenomena needing consideration. I earn the right to generalize only to the degree I have penetrated the specific phenomena from which I am generalizing.

Of course, if I keep my comments as observationally faithful as possible, they are likely to contain at least *some* truth, just as a person uneducated in art can, if conscientious and careful in observation, say something more or less valid about a painting. But we need to keep in mind that the distance from *some* truth to *profound* truth is likely to be huge. To speak as I did about an "integral symphony of the morning" places a huge burden on me to understand, in depth, not only this flower but also almost everything else in and under the sky. I have not even begun carrying out such a task here, having done no more than suggest a few things one might attend to.

There is, however, one consolation in all this. While it's true that talking about "the character of the morning" raises almost impossibly high expectations, the fact is that we are making the *same kind of judgment* when we come to appreciate the expressive qualities of one particular flower—and even, in however minimal and unconscious a way, when we recognize from its expressive character that the flower is of this species rather than that (Brady 2002). Moreover, because we are always dealing with interpenetrating unities, we find our understanding of the flower leading us in ever-expanding contextual circles to the qualities of the largest whole we can encompass with our perception and thought. It is one of the features of integral wholes that every part is an expression and revelation of the whole (Bortoft 1996, part 1).

Light, Conversation, and Joy

I hope now that two brief, additional thoughts about the morning will not seem unduly strange. A sunrise is often felt, not only as a glorious, radiant event, but also as the occasion for an outburst of joy. If you consider the ascending, radiating, and expansive gestures mentioned above, you will recognize that they are somehow resonant with human joy. All you need to do to verify this is to pronounce twice and in a heartfelt way the sentence, "I am so happy!"—once while swelling your chest, letting your arms expand upward and outward, and moving with your gaze into the surrounding environment, and once while contracting your limbs, body, and consciousness toward your center. In the latter case you will immediately recognize the grotesque inappropriateness of the movement, while in the other the movement seems perfectly natural, if not inevitable.

This is not to say that we should naively ascribe our human joys to nature. Nothing could be more foolish. But there *is* an objective, demonstrable connection between our joy and certain physical gestures, and this can be the case only because the physical gestures have an inner, speaking content.

My second observation has to do with that morning chorus of birds contributing so powerfully to our sense of joy. A friend of mine once remarked, "The sun rises, and the birds feel compelled to sing the light." This puzzled me until I understood a certain gesture characteristic of all meaningful sound. Sound moves outward, like ripples from a stone thrown into a pond. But, again, I'm not referring primarily to the physical movement of sound waves.

Often, if we're wondering about the qualities of something, we can find initial guidance by turning our attention to the wisdom inherent in our everyday language. We are frequently *struck* by someone's words; even when spoken softly, words may *slap* us in the face or *hit* us in the gut. In other cases, the words may more subtly penetrate us, insinuating themselves into our subconscious. The cartoonist, with typical exaggeration, depicts someone being "blown over" by a shout, or he might indicate speech (or the song of a bird) by drawing lines radiating from the creature's mouth just as a child draws lines radiating from the sun. If you try to imagine a movement in the opposite direction, you will immediately recognize that it does not fit.

Speech is undeniably an expansive and radiating, light-like phenomenon. Without light, the world is not there for us with any clarity. But without the conceptual illumination of the word, the world is also not there. We cannot see what we have not learned to discriminate through the conceptual power of the word. The word, perhaps we could say, is the inner being or essential meaning of the light—an idea, I imagine, that might prove fruitful for the physics of the future. Even now the physicist investigating light at the quantum level is inclined to say that the experimenter's conversation with the light somehow shapes its manifestation.

Just as speech is light-like, so, too, the raying light has from ancient times been understood as speech-like. In the Upanishads it is recorded that "the Sun is sound; therefore they say of the Sun, 'He proceeds resounding.'" Ananda Coomaraswamy, drawing on the ancient texts of the east, summarizes the matter this way: "The shining of the Supernal Sun is then as much an 'utterance' as a 'raying'; he, indeed 'speaks,'" and "the Sun himself 'sings' as much as he 'shines'" (1977, 153 n, 193).

Again, our routine language is suggestive of the intimate relation between light and speech. We refer to "bright" ideas and "brilliant" sayings, and we respond "I see" when we have understood someone's words. If we attend with any sensitivity to our actual meaning when we say these things—meaning that often arises from genuine perception at some level of our being—we can begin to appreciate the delicate interweaving of light and speech. Goethe seems to have been honoring this intimate connection when he wrote the following bit of dialogue in his fairy tale, "The Green Snake and the Beautiful Lily":

> "What is more glorious than gold?" asked the king.
> "Light," answered the snake.
> "What is more refreshing than light"? asked the former.
> "Conversation," replied the latter.

I hope all this not only suggests a possible truth in my friend's characterization of the birds as "singing the light," but also may save us from the kind of arrogance at work when, hearing the poet or prophet or nature-lover say, "Heaviness may endure for a night, but joy cometh in the morning," we dismiss these words as unscientific, merely subjective sentiment. Certainly our experience of joy may commonly be combined with purely personal elements, but this leaves open the question of whether there is an objective character of joy not only at the root of our own experience but also displayed in nature herself.

A More Difficult Objectivity

There is a well-known school of acting, first articulated by the Russian actor and director Michael Chekhov in the first half of the twentieth century, based on the objective character of physical forms and movements. The actor, faced with the need to express, say, pride or mortification or joy, does not attempt to summon from memory a prideful or mortifying or joyful episode from his own past so as to re-enter that earlier personal context and psychological condition. Rather, he finds gestures in the world from which he can draw directly the inner expressive character he is looking for. I have seen an actor, purely as an exercise, almost instantaneously transform his eyes into an expression of profoundest, tearful grief, while disavowing any sort of *merely personal* feeling. If, as I suggested earlier, the experience of joy naturally embodies itself in a certain expansive gesture, so, too, the outward gesture naturally takes form inwardly as an experience of joy.

Through this experience we come to know something about the world. As Vladislav Rozentuller has written,

> To move your hand toward an object in a certain hesitating and faltering way is (for the actor whose powers of perception and attention have been trained) to experience in the quality of the movement a feeling of distracted worry or anxiety. The feeling is objective in the sense that it belongs to the physical movement itself; the actor need not recall or imagine any purely personal anxiety. But, at the same time, the feeling does become *his* feeling. We could say that the experience has a subjective-objective character: the actor makes of his personal consciousness a stage onto which he invites this or that feeling from the objective world. (Rozentuller and Talbott 2005)

The actor onstage cannot help realizing that the world is word-like; every outer form corresponds to an inner content, and every inner imagination can be given its most natural outer form. A gesture can be grotesquely inapt or powerfully revealing—a simple fact that testifies to the significance, or speaking quality, of the world's forms. The kind of training undergone by actors of the Chekhov school strikes me as very like one kind of training required by the practitioners of a new, qualitative science. The only way to recognize the wholeness of nature in all its expressive power is to perceive it with the full range of expressive powers of the human being. The instrument of perception must be equal to its object. We will never develop a truly holistic science as long as the scientist must paralyze or imprison major human capacities—for example, the capacity to recognize the very real unity of a great work of art.

When we accept the human being as the primary instrument of scientific understanding—when we realize that we must discover within our own powers of speech what speaks in the world—then the need for uncommon inner discipline becomes apparent. This is what Owen Barfield had in mind when he wondered why there is any need "to make quite such a song and dance" about objectivity in the more usual sense (1977, 139). After all, it shouldn't be so hard to get rid of personal bias if there is no genuine personal connection between ourselves and the things we're investigating. "To put it rudely," Barfield expostulated, "any reasonably honest fool can be objective about objects." But it's altogether different when we must attend "not alone to matter, but to spirit; when a man would have to practice distinguishing what *in* himself comes solely from his private personality—memories, for instance, and all the horseplay, of the Freudian subconscious—from what comes also from elsewhere. Then indeed objectivity is not something that was handed us on a plate once and for all by Descartes, but something that would really have to be *achieved,* and which must require for its achievement, not only exceptional mental concentration but other efforts and qualities, including moral ones, as well."

Indeed, the task may have been too great for humankind to attempt at the dawn of modern science. We can imagine there was a deep, unconscious wisdom in the resolve to shackle the greater part of the human instrument and subject ourselves to the discipline of mathematics, where a certain kind of rigor and objectivity *are* almost "handed us on a plate." Without that preliminary training, it would have been nearly impossible to subdue the disorderly babel of voices still reigning in the human soul—voices of magic and superstition, of myth and legend, of religion and irreligion, of ethnic pride and prejudice—voices still capable of disrupting in childish ways the sober, geometric imaginations of Kepler, Galileo, and even Newton.

But we have completed this training—more than completed it, for we have carried our mathematization of reality to the unhappy point were the world begins to disappear behind a ghostly veil of abstraction. This veil conceals the perceptible, testable world from us as effectively as the old metaphysics ever did. Today, if we would test the phenomena around us, we have the opportunity to bring to them not only our measuring rods and mechanical instruments, but our full-fleshed capacity to speak the living language of the phenomena, a capacity now chastened by our awareness that "even where we do not venture to apply mathematics we must always work as though we had to satisfy the strictest of geometricians" (Goethe 1995, 11–17).

We do not, after all, have to accept a science lacking in rigor. We only need to realize that there are two different, almost opposite ways to seek ideal clarity and precision. One is by following the path we traced earlier, admitting into our science only what we can grasp unambiguously, only what we can lay hold of, immobilize, and tie down, only what can be isolated as a separate thing and analyzed strictly in terms of its external or mechanical relations with other isolated things. In such a spirit (rudely disturbed by the discoveries of the past century), physicists have always sought for "fundamental particles"—particles lacking in qualities and accounting for the world's phenomena solely through their aggregate configurations, that is, solely through their clean, mathematically describable, external relations.

We gain a very different kind of clarity, not by minimizing the qualitative, phenomenal content of our scientific descriptions, but by maximizing it. We illuminate a phenomenon from every possible side, in every different light, exploring its contextual relations and potential for transformation as fully as we can. This clarity is not attained by stripping reality down to a formal grammar. It's the clarity produced by fullness of understanding rather than ease or simplicity of understanding. Instead of obscuring phenomena with the blinding white light of abstraction, and so reducing them to a kind of black-and-white skeletal syntax, we open ourselves to receive the phenomena in all their full-throated color.

Then, perhaps, it will not be too much to hope that we as scientists may learn to "sing the light" of creation—not as voyeurs staring at a cold and alien world disconnected from our own life, but as participants in a new morning of creation when, if we make ourselves worthy instruments, the Word will rise up in us as a song of understanding.

Chapter 14

Delicate Empiricism

Practicing a Goethean Approach to Science

I (Craig Holdrege) have vivid memories of Mr. Sinn's ninth grade science class. We did experiments with glassware, tubes, and Bunsen burners—they were fun. But then Mr. Sinn taught us how to explain the results of our experiments. He described molecular processes that we didn't see. These became schemes with letters and numbers on the blackboard. We now were supposed to know what had really been going on. And I was lost. I didn't get it. What did the blackboard diagram have to do with what we'd been observing? This was an unsettling experience that had significant consequences: I avoided science like the plague in high school.

It is also my first memory of the kind of experience that I have had repeatedly since then, and that has been key in my pursuing Goethe's approach to science. It is the experience of confronting what are called scientific explanations, and feeling (in thought) a distinct sense of dissatisfaction. How can a phenomenon be explained by something that is supposed to underlie it and that is always less than the phenomenon itself? I have been amazed that what a large community of people feels to be an explanation leaves me with the question: What do I have now? What am I doing by leaving the phenomenon in order to explain it? Let me give a few examples.

In a college botany course I learned why plants that grow in shady places have broader and larger leaves than plants that grow in full sunlight. The reason given is that plants growing in shade don't receive as much light to do photosynthesis. Therefore they grow larger surfaces with which they can capture more light and produce more organic matter via photosynthesis. Plants have developed this strategy to survive and reproduce in shady habitats. This is a typical functional explanation that makes perfect sense—until you think the matter through a bit further. The larger the surface area a plant creates, the more substance it needs to build up and sustain its larger body. Wouldn't it be just as effective for the plant to stay very small with narrow leaves? In this way it wouldn't have to do so much photosynthesis since it could stay small. Both explanations make sense. I have yet to find a functional explanation of a phenomenon for which one couldn't find equally plausible alternatives. Such evolutionary explanations always fall short. They fall short because they are an attempt to get a grip on a complex biological phenomenon from only one narrow and limited perspective. I have shown this to be the case in celebrated textbook examples such as the long-necked giraffe and industrial melanism in the peppered moth (Holdrege 2004b, 2005b).

I can formulate the problem in another way. Every biology student learns that the fundamental question of biologists confronting a phenomenon is: What is the underlying mechanism? It may be a Darwinian survival strategy, or a hormonal or genetic mechanism. In the search for such mechanisms two essential things happen. First, you isolate the phenomenon out of its context within the organism as a whole, and second, you seek to explain it in terms of a reduced set of quasi-mechanical processes. In the end what you come up with is a simplified picture of a phenomenon caused by an abstractly conceived underlying mechanism. (The neurologist Kurt Goldstein has elucidated this problematic side of science in his seminal work on holistic science, *The Organism* [1995].)

Take, for example, a "trait" that is explained by a "gene." When Mendel discovered hereditary patterns in pea plants, he focused his attention on particular characteristics, such as seed shape or flower color. (For an in-depth discussion see Holdrege 1996.) He mentally abstracted these characteristics from the plant. He could only do this with "clear and distinct" characteristics, that is, ones that don't vary much under changing conditions. He did not look at flower color as it changes from bud formation through flower maturation and wilting, nor did he worry about the slight variations in color that occur between different specimens of the same breeding line. Flower color grasped as a Mendelian trait entails bracketing out developmental changes and variation. The resultant trait is an isolated, distinct feature, and it is quite straightforward to go from it to an underlying particulate factor—later called a gene—that is inherited and responsible for the appearance of the trait through the generations. The history of genetics shows the power of this way of viewing and working with organisms.

The problem is that both the trait and the gene are products of abstraction, so that one is explaining an abstraction with an even greater abstraction. The red flower color of a strain of pea plants is much more than a genetic trait, and the biochemical component of inheritance is much more than the genetic code. This, as we have seen in previous chapters, is becoming increasingly clear even within the field of genetics, so that some geneticists question the value of the concept of the gene altogether. It's interesting how reality tends to catch up with scientific abstractions at some point.

Over the past twenty-five years I've become keenly aware of what scientific concepts do not tell us and do not reveal. And because they are often used as if they told us something beyond their own narrowly circumscribed domain, they often mislead and cover up phenomena. Becoming aware of such boundaries is significant, because—to paraphrase Hegel—in our gaining an awareness of a boundary we have already begun to transcend it. We are at the gateway to a new kind of understanding, to the further evolution of science that Goethe inaugurated.

Goethe and Modern Science

> There is a delicate empiricism that makes itself utterly identical with the object, thereby becoming true theory. But this enhancement of our mental powers belongs to a highly evolved age.*

The German poet and scientist Johann Wolfgang von Goethe wrote these words at the age of eighty, just three years before his death in 1832. They are his mature articulation of fifty years' work to establish an explicitly participatory, phenomena-based scientific practice. As a poet and writer, Goethe had attained fame and recognition already when he was in his twenties. Today he still looms as a creative giant in the history of German literature. Far fewer people have recognized the value of his work in science.

* Written by Goethe (1749–1832) in 1829 and included in the chapter "Reflections in the Spirit of the Wanderer" in his novel *Wilhelm Meister's Journeyman Years.*

While some of Goethe's scientific discoveries have entered mainstream science, his methodology has not. And yet Goethe's approach has also never been fully ignored. Repeatedly it has been the focus of vibrant discussion about the nature of science. Many great minds of modern science have found it necessary to grapple with Goethe the scientist—Darwin, Haeckel, Helmholz, Sherrington, and Heisenberg, to name a few. Thousands of scholarly articles and many volumes have addressed Goethe's approach. (See Amrine 1987 and, for more recent publications, http://www.natureinstitute.org/resources.) And yet, Goethe is a perennial outsider. I think it is fair to say that within the broader contemporary scientific community his efforts are virtually unknown or deemed irrelevant to the advancement of science.

But the very fact that Goethe's way of doing science ever and again becomes a topic of discussion suggests that he may have hit a central nerve concerning the problems and tasks of human knowledge. When we become aware of the boundaries and limitations of the conventional scientific approach and search for orientation, Goethe's work remains a bright and unique source of light illuminating a pathway into new terrain.

Although Goethe is often portrayed in opposition to science, he viewed his efforts as a further refinement of scientific method. What has made this Goethe-inspired evolution of science both enticing and forbidding is that it involves, in Frederick Amrine's words, "the metamorphosis of the scientist" (Amrine 1998). Goethe knew that his delicate empiricism entailed "an enhancement of our mental powers," and for that very reason it still remains in its infancy. It entails becoming aware of the "object" view of the world that so strongly informs both our everyday and scientific thinking. When we leave this "natural attitude" (Husserl 1962, 39) behind, we can begin to see how we participate within the world and then work to gain new bearings for our thinking and perceiving. This is the path—both arduous and exhilarating—that Goethe trod. This chapter describes an individual journey into this new terrain.

Delicate Empiricism: Science as a Conversation

The realization that the phenomena we confront are always richer than the abstractions we use to explain them is central to a Goethean approach. This realization is the expression of a twofold awareness or sensitivity that Goethe points to with his expression "delicate empiricism." First, we experience a phenomenon (a mouse, a wooded swamp, a range of blue hills in the distance, or the clouds moving across the sky) as a kind of fullness that calls forth wonder, curiosity, questioning. We want to get to know it better, or as Goethe states it radically, "become utterly identical with it." This is empiricism, because we orient all our striving around the phenomena themselves. A phenomenon is what meets the eye, but we also experience it is as something more, as a kind of surface that is pregnant with a depth we may be able to plumb. But we realize that we will not fathom these depths with models and theories, which more likely than not will lead us away from the phenomenon itself. This brings us to the second mode of sensitivity: we are acutely aware of the thoughts we bring to the phenomenon, how we interact with the world through thinking. We know that in conceiving thoughts we can both illuminate and color our experience. The more we are aware of the thoughts we bring, the more transparent and illuminating they can be. We must become delicate in the way we work with our concepts in our efforts to let the depths of the phenomena disclose themselves.

Goethe describes the process of gaining knowledge in the following way:

> When in the exercise of his powers of observation man undertakes to confront the world of nature, he will at first experience a tremendous compulsion to bring what he finds there under his control. Before long, however, these objects will thrust themselves upon him with such force that he, in turn, must feel the obligation to acknowledge their power and pay homage to their effects. When this mutual interaction becomes evident he will make a discovery which, in a double sense, is limitless; among the objects he will find many different forms of existence and modes of change, a variety of relationships livingly interwoven; in himself, on the other hand, a potential for infinite growth through constant adaptation of his sensibilities and judgment to new ways of acquiring knowledge and responding with action. (Goethe 1995, 61, written in 1807)

In Goethe's view science entails "mutual interaction" with the phenomena. Engaging in this process, we discover the "limitless" nature of connections and relationships in the world, but at the same time we grow and adapt ourselves to new, more adequate ways of knowing. Doing Goethean science means treading a path of conscious development. The question accompanying every aspect of the work is, "How can I make myself into a better, more transparent instrument of knowing?" In traditional science, we are much more likely to ask, "How can I find ways of adapting the phenomena to my specific approach so that I can answer my question?"

I have found the metaphor of conversation increasingly helpful in illuminating the nature of a Goethean approach to science. The metaphor brings to consciousness that doing science is a back-and-forth between partners in an ongoing process. It accentuates a kind of inner attitude that lies at the heart of doing Goethean science, one very different from the frame of mind one normally associates with science (although it informs, but often not explicitly, the work of many good scientists). Here, expressed in fairly general terms, are some of the elements of science-as-conversation. (See also Talbott 2003a.)

1. When I enter into a conversation with nature, my interest has been sparked by some experience, my attention has been caught. I'm presented with a riddle and begin asking questions, observing, and pondering. In this way I give the conversation an initial focus. If the interaction between me and nature has no focus it can easily become chit-chat and not a conversation.

2. But if the focus I bring is too narrow and too rigid (for example, a narrowly defined hypothesis), we don't have a conversation; we have a drill (one-sided questioning). In any productive conversation the process itself is paramount. It's not just about me answering my preformulated questions, but centrally about what happens along the way. There will be surprises, moments of silence, tension. The back and forth between me and nature is dynamic, and I attend to this process as an integral part of the conversation.

3. Taking the conversation-as-process seriously means realizing that it is open-ended. I don't know where we're going to arrive. At every moment the conversation is imbued with an atmosphere of openness. I could also describe this attitude as a kind of animated looking forward to the unexpected.

4. Nature is my partner in the conversation. If I truly mean this and don't take the statement as a feel-good cliché, then I'm acknowledging that nature is something in its own right. I may not, at the outset, be able to say more than that. But the recognition of the other as something

in its own right is a pre-condition for any conversation. This recognition infuses respect into the conversation and gives it dignity. In saying this I don't mean that geologists will no longer crack open rocks with their hammers or botanists will stop pressing plants. However, knowing that I am involved in a conversation makes me more circumspect and I become more sensitive in what I think and do. I may ask myself, for example, whether I may be going too far and transgressing boundaries. I'm not talking here about abstract, prescriptive directives—since the conversation *is* a process, I can't know what will emerge out of it beforehand. But in any case, it is carried by an attitude of respect.

5. An essential feature of the conversation is that I listen to what nature has to say. Receptive attentiveness allows us to hear and see with fresh ears and eyes. It's the quality of open interest in what the other has to say. But it would not be a conversation if I only listened. I respond and interject. I am actively giving form to the conversation through my questions, observations, and the new concepts I bring in. A vibrant conversation needs the movement between receptive attentiveness and active contributing.

6. In the course of any real conversation the partners change and evolve—they are in a different place than they were at the outset. It is easy to see that I as a scientist change in this conversation. I gain new experiences, take new qualities into myself, and get to know the world more deeply. And in a simple sense, any time we interact with nature through an experiment, we change nature. Field ecologists have recently discovered that even touching and marking plants in the field can affect their growth (Cahill et al. 2001). Goethe's seminal essay "The Experiment as Mediator between Object and Subject" (Goethe 1995, 11–17, written in 1792) shows his keen awareness of science as a way of interacting with nature. Experiments don't "prove," they mediate a relationship—they are a means of engaging in a conversation. We are interwoven with nature and weaving a new fabric when we do science.

But there is more. When I do plant studies with participants in courses, we spend a good deal of time carefully observing individual plants, building up a shared picture of their characteristics. We also speak about the processes we go through. We all recognize how the plant has changed us. But it is another matter when I ask about the plant: is it any different? Certainly not if we are thinking abstractly about its molecules or about its outer form or structure. But if we stay descriptive and simply acknowledge what has taken place, we can say: through our engaging in observation and contemplation of the plant, the plant appears in a new form, namely in human consciousness. Nature finds a new expression through the process of human knowing. This participatory relation between the knower and the known, a relation that truly involves both partners, was most clearly stated by Rudolf Steiner, who based his epistemology on his decade-long work as the first editor of Goethe's scientific writings. As Steiner, writing in 1894, puts it:

> Does not the world produce thinking in human heads with the same necessity as it produces the blossom on a plant? Plant a seed in the earth. It puts forth root and stem; it unfolds into leaves and blossoms. Place the plant before yourself. It connects itself, in your mind, with a definite concept. Why should this concept belong any less to the whole plant than leaf and blossom? You say the leaves and blossoms exist quite apart from a perceiving subject, but the concept appears only when a human being confronts the plant. Quite so. But leaves and blossoms also appear on the plant only if there is soil in which the seed can be planted, and light and air in which the leaves and blossoms can unfold. Just so the concept of a plant arises when a thinking consciousness approaches the plant. (Steiner 1964, 64–65, translation slightly modified by Craig Holdrege)

This may seem to be a radical idea. It hits up against deep-seated assumptions we have about our relation to the world. Most of us are held captive by the notion of the world "out there," separate from us "in here." The moment we wake up to the fact that we are part of the world and engaging in a conversation with her to get to know her (and ourselves) better, the captivity of a dualistic world view ends. This is, as I see it, one of the most important implications of a Goethean epistemology. We are freed to engage as participants in *one* world.

7. The realization of our participation helps us to see one more facet of science-as-conversation: I become aware that I am taking on a responsibility. I'm engaging in the world, and whatever the outcome of the conversation, it will bear in part my stamp. I put to rest once and for all the comfortable specter of something called "value-free" science engaged in by some detached being called a scientist. Science is all about participation, and I can't distance myself from the process and its results.

So much for an introductory overview. The idea of science as a conversation grows out of the doing. But once you've become conscious of it, it becomes a kind of scientific conscience—an inner guide—for all further work: Am I aware enough of the process? Is a back-and-forth occurring? Am I listening, or pushing an agenda? When your work becomes infused with a circumspect attitude of questioning wedded to a strong desire to engage in the phenomena, you can see what Goethe wanted to express with the phrase "delicate empiricism." And you can also understand why he added that its practice belongs to a "highly evolved age," since it is dependent on transformation within the human being. Goethe's science involves the consciously evolving scientist.

In the following sections of this chapter I will try to illuminate more fully the process of doing Goethean science by way of an example. I'll show how science-as-conversation can unfold and also discuss additional important features of a Goethean approach.

Engaging the Conversation

When I moved to the Northeast twelve years ago I met new habitats, plants, and animals. Early in March I was down in a wooded swamp and saw some strange looking plants—maroon and yellow, fist-sized buds that emerged directly out of the icy ground. They had a beautifully curved and pointed form, something like the hats of elves you see in children's books. Nothing else in the wetland showed any sign of spring in this gray, frozen world. I was captivated—I began my journey to get to know the skunk cabbage.

So that's how it begins. Something captivates your interest, and you move toward it. For me this meant returning to the skunk cabbage again and again—in all seasons and at different times of the year. I did this over a period of six years, in which time I also read everything I could get my hands on regarding skunk cabbage (which wasn't a whole lot). What was my purpose? What was my goal? I know of no other answer than to say: I wanted to get to know the skunk cabbage. I felt it as a riddle that drew me toward it.

I didn't have a particular hypothesis that I wanted to test. I didn't want to "explain" the plant or its features in terms of competition or survival. Since I had been practicing the Goethean approach for many years, it wasn't very hard to avoid the trap and narrowing effect of wanting to explain. But when I'd go out with other people, I'd often be asked questions such as: Why does it flower so early? Why does it heat up? Why do its leaves grow so large? I could tell them what some scientists thought, and perhaps point out alternative explanations. But I'd also say, and this is more important, that before we can see whether it's even meaningful to ask such questions, we have to get to know the plant much more intimately. And those very same "why" questions can hinder us from doing so.

So the conversation began. I began building up a picture of the plant's development through the year. To do this I made lots of detailed observations. The plant itself is a unity that transforms over time. I had a vague sense of that unity, but I had to get to know it by bringing together discrete observations. As Henri Bortoft puts it, "the way to the whole is into and through the parts" (Bortoft 1996, 12). In every part you discover new phenomena, and new questions arise. One danger here is that you let yourself get pulled into an endless process of analysis, where the whole and, successively, each part dies into further analysis. The only antidote I know to this problem—which is a major problem for all of science—is to periodically disengage from analysis, step back, and ask yourself: How does this all relate to the skunk cabbage? I began my journey wanting to get to know the plant better. So I have to continually try to place all the knowledge I gain through engaging in the parts (analysis) back into the context of the plant as a whole (synthesis).

I tried to go down to the wetland to observe every week or two. I sketched the plants to help me look more carefully, and I also took photographs. Sometimes I would have specific questions: How are the flowers really shaped? Are the skunk cabbages that grow at the wetland edge any different from those that grow in the wetter core area? But at other times I would purposely go out without a content focus, with the attitude "Let's see what comes today." I know some of my most interesting observations—such as discovering that bees on one of their first outings of the season were visiting skunk cabbage flowers—came when I was walking with an "unframed mind."

Through the weaving interplay of open awareness and focused observation we come to know the phenomena. It's the most time-consuming part of doing science and proves most fruitful when you can move back and forth between the poles of focus and opening outward.

For most of us, focusing is easier than observing with an unframed mind. Of course, we never engage in observing with a fully unframed or unfocused mind. We always have some kind of intentionality or attentiveness that orients us toward the world (or we fall asleep). But we *can* work on developing a kind of open, listening awareness. As Thoreau writes, "Be not preoccupied with looking. Go not to the object, let it come to you. . . . What I need is not to look at all—but a true sauntering of the eye" (September 13, 1852, Journal 5: 343–44, quoted in Walls 1999, 46–47).

Sometimes immersing oneself in new impressions can help transform habitual ways of thinking and open our eyes to qualities. Having worked as a minister in the Dukedom of Weimar for ten years, gained fame as a writer and poet, and already carried out an array of scientific studies, Goethe felt stifled. He needed a change and took a radical step. After celebrating, on August 28, 1786, his thirty-seventh birthday with friends in the spa city of Carlsbad, Germany, he stole away in the middle of the night, incognito on a postal coach. His goal was Italy. He crossed over the Alps and arrived in northern Italy, where he wrote in his journal: "At present I am preoccupied with sense-impressions to which no book or picture can do justice. The truth is that, in putting my powers of observation to the test, I have found a new interest in life. How far will my scientific and general knowledge take me: Can I learn to look at things with clear, fresh eyes? How much can I take in at a single glance? Can the grooves of old mental habits be effaced? This is what I am trying to discover" (Goethe 1982, 21, written September 11, 1786).

This journey did in fact help Goethe get out of old grooves and see with fresh eyes. He gained key insights into the nature of plant metamorphosis while immersing himself in the new and fascinating vegetation of Italy. When an actively open and searching gaze engages phenomena as a fresh presence, they can begin to speak in new ways.

Exact Sensorial Imagination and Living Understanding

After I go out and observe, I make a point of actively remembering the observations. With my mind's eye I inwardly recreate the form of the leaves, I inwardly sense the colors and the smells, and so on. This process of conscious picture building is what Goethe called "exact sensorial imagination" (Goethe 1995, 46, written in 1824). It entails using the faculty of imagination to experience more vividly what I have observed. I try to be as precise as possible—and will often notice where I haven't observed carefully enough, which I try to do the next time I'm out. When you do this kind of conscious picture building, you grow more and more connected to what you're observing.

But there's something else. The plant begins to reveal itself as a process. When we begin observing, we have many separate images, and we have to overcome separateness to begin seeing the plant as the living creature it is. The life of a plant plays itself out in the ongoing unfolding and decay of organs (leaves, stalks, flowers, etc.). We are presented with a drama of transformation that we can enter into. But we can't enter into it through observation alone. We need to utilize our faculty of imagination to connect within ourselves what is already connected within the plant. As Goethe writes: "If I look at the created object, inquire into its creation, and follow this process back as far as I can, I will find a series of steps. Since these are not actually seen together before me, I must visualize them in my memory so that they form a certain ideal whole. At first I will tend to think in terms of steps, but nature leaves no gaps, and thus, in the end, I will have to see this progression of uninterrupted activity as a whole. I can do so by dissolving the particular without destroying the impression" (Goethe 1995, 75, written in 1795).

So to begin to grasp the flow of life and its specific qualities in skunk cabbage you have to work to make your thinking fluid (process-oriented) and dynamic. In Goethe's words, "If we want to approach a living perception (*Anschauung*) of nature, we must follow her example and become as mobile and flexible as nature herself" (Goethe 1995, 64, this translation by Craig Holdrege).

I'd now like to give what you might call a report on my conversation with skunk cabbage. But it actually wants to be more than that. It's an attempt to give a portrayal, to paint a picture in words that will let you see something of the unique qualities of this plant. I hope to give you a glimpse of another being, although I'm all too aware of my inability to adequately express what I have met.

Skunk Cabbage—A Portrayal

To find the first spring plant in flower in our region—the edge of the Taconic range southeast of Albany, New York—you have to get out before it feels much like spring at all. It's March, the ground is still frozen, and frost comes nearly every night. Walking through the woods down a soft slope, you see the gray and brown tree trunks, a coloring mirrored in the ground litter of leaves from the previous year. There is no green. Not only the temperature but the whole mood of the woods is cool.

At the base of the slope there is a wooded wetland—a flat expanse in which patches of ice spread around islands of bushes and small trees. In this still, quiescent world, little centers of emerging life are visible, the first sign of early spring—four- to six-inch-high, hoodlike leaves that enclose the flowers of skunk cabbage. (See figures 14.1 and 14.2.)

Figure 14.1 *(above)*. A group of skunk cabbage spathes and leaf buds in March. (Drawing by Craig Holdrege)

Figure 14.2 *(left)*. Skunk cabbage spathes. (Drawing by Craig Holdrege)

Both color and shape are striking. Some leaves are deep wine-red or maroon, while in others this background coloring is mottled with dots or stripes of yellow or yellow green. The shape is like a spiral, sculpted hood that leaves only a narrow opening on one side. Not only the colors, but also the specific shapes are manifold; some are pointed and strongly twisted, others rounder and squat. As my eye sweeps over the twenty or thirty plants before me, my gaze is brought into a spiraling movement when it tries to rest upon any single specimen. The deep color is warm, the sculpted form alive.

Looking at skunk cabbage on one of the first warm, sunny March afternoons (it's maybe 50 degrees Fahrenheit) with the light shining through the leafless trees and shrubs and illuminating the wetland floor, I often sense for the first time that spring is on its way. On such days I've even seen the first bees of the year flying in and out of the skunk cabbage hoods.

The hood is, in botanical terms, a highly modified leaf called a spathe. The spathe wraps around itself to form a space that encloses a spherical head of flowers, called a spadix (see figure 14.2). The spathe functions as a bud that holds and protects the flower head when it emerges out of the ground. But it is a bud that never unfolds. When the flowers are full in bloom, they are still enwrapped by the spathe. You can see the flower head only by peeking inside the narrow opening in the spathe.

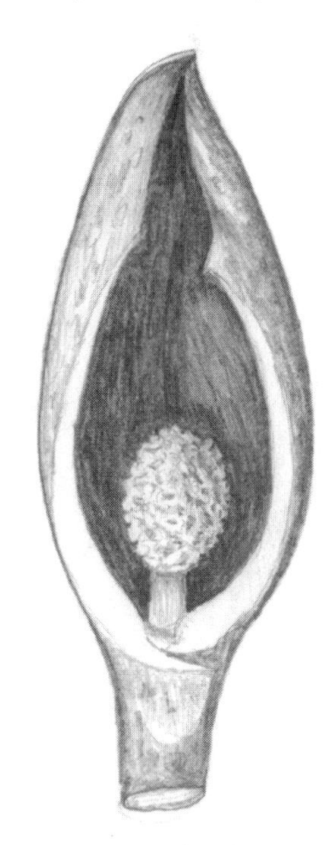

Figure 14.3. Skunk cabbage spathe; the front part has been cut off to show the flower head (spadix). (Drawing by Craig Holdrege)

The roundish flower head (about 2 centimeters in diameter) has a spongy consistency like the spathe itself. It consists of numerous small, tightly packed individual flowers (see figure 14.3). They have no petals, which make up the showy part of the flower in most plants. Rather, they have four inconspicuous, fleshy, straw-colored sepals (which in many plants form the bud leaves enclosing the petals) that never really unfold.

The flowers "bloom" when the stamens grow up between and above the sepals and release their pale yellow pollen. Following this the style grows out of the middle of each flower to be pollinated by insects carrying pollen from other flower heads. All of this happens *within* the enclosing spathe. These first flowers of spring never leave their protective enclosure.

A couple of times I've been lucky enough to see spathes growing up through a thin layer of ice, with ice melting around the spathe in a circular form. This is an indication of skunk cabbage's remarkable capacity to produce heat when flowering. If you catch the right time, you can put your finger into the cavity formed by the spathe and feel your fingertip warm up noticeably when you touch the flower head. I have measured the temperature at the base of the flower head numerous times, and have found it to be as warm as a 61 degrees Fahrenheit when the surrounding air temperature was only 32 degrees Fahrenheit. Biologist Roger Knutson found that skunk cabbage flowers produce warmth over a period of twelve to fourteen days, remaining on average 36 degrees Fahrenheit (20 degrees Celsius) *above* the outside air temperature, whether during the day or night (Knutson 1974). During this time they regulate their warmth, as a warm-blooded animal might!

Physiologically, the warmth is created by the flower head's breaking down substances while using a good deal of oxygen. The rootstock and roots store large amounts of starch and are the likely source of nutrients for this breakdown. The more warmth produced, the more substances and oxygen consumed. Knutson found that the amount of oxygen consumed is similar to that of a small mammal of comparable size.

We must imagine that as the spathe grows out of the usually frozen ground, the flower head heats up and the warmth radiates outward. While in this heating phase, the flowers bloom, releasing pollen and being pollinated by insects. Not only can you see the first insects flying around between skunk cabbages, but you also find beetles and spiders crawling around within the warm enclosures of the spathes. You can even discover a spathe opening veiled with a spiderweb.

The flowers also release a noticeable odor at this time. On a calm day coming down to the wetland you can smell a lightly pungent, somewhat skunk-like odor. If you put your nose to the opening of a spathe or break off a small piece and crush it between your fingers, the scent is markedly stronger. Small flies and other insects are attracted to the flowers by the smell.

Due to the warmth production, a constant circulation of air in and out of the spathe occurs. From the flower head, warmth is generated and the air moves up and outward, while cooler air is drawn into the spathe. A vortex is formed, with air streaming along the sculpted, curved surfaces of the spathe. In a habitat with numerous skunk cabbages, a microcosm of flowing warmth and odiferous air is created in which the first insects of spring fly.

This is the world of skunk cabbage over a number of weeks in March and sometimes into April: on the one hand, the enclosed, protected life just peering out of the still wintry earth, and a flower that remains in a bud; on the other hand, the active, warmth-, movement-, and scent-emanating organism that creates a unique environment for the first stirrings of insect life. So skunk cabbage blooms at ground level in a bud that doesn't open, while at the same time helping to create the environment for its own development and that of other species.

When the spathe emerges out of the ground, there is often the tip of a large bud next to it, sticking an inch or two out of the ground (see figure 14.4). This bud contains all the leaves that will develop on the plant and is often already visible in the previous fall, having developed in the summer and overwintered. Only when the spathe slowly begins to wilt does this tightly packed bud of leaves begin to grow. It grows longer than the spathe and is shaped like the tip of a spear. Then, when the days begin to get noticeably warmer at the end of April and into May, the bud unfolds rapidly. It's clear that skunk cabbage now needs outer warmth to develop. The bright green leaves unfold in a beautiful spiraling pattern. Each leaf is rolled in upon itself and at the same time enwraps the next leaf. It's the closest thing to an archetypal process of unfolding you can imagine.

Gradually a large, funnel-shaped rosette of leaves forms. The largest leaves reach three, occasionally four feet in length. By mid-May this surge of growth peaks and the wetland is flooded with green patches of skunk cabbage. The leaves are oblong in shape and have a long leaf stalk. The leaf stalks are thick, but also easy to crush. They have no woody fibers and consist primarily of air and water inlaid with soft plant matter. This consistency extends, untypically, into the flowering part of the plant: both spathe and flower are watery and spongy. By contrast, think of the distinct difference you find in a wild rose between the hard, prickly, woody stems carrying divided, fibrous leaves on the one hand, and the refined, almost rarified petals on the other.

Figure 14.4. The development of skunk cabbage from early spring to July, when its leaves begin to disintegrate. (Drawing by Craig Holdrege)

A crushed leaf also exudes a skunk-like odor, and ingested leaf juice calls forth a strong inflammatory reaction in the mouth and esophagus of humans. Skunk cabbage not only produces its own warmth; it also stimulates warmth processes in us. Few creatures eat the leaves. I've seen leaf buds and also spathes that have been nibbled upon. In one instance the wetland was covered with a late March snow and tracks of wild turkeys led up to the buds, which apparently they had eaten from. Early Swedish settlers in Pennsylvania gave skunk cabbage the name "bear-weed," since bears were known to feast on the buds and leaves.

In our area the leaves of the trees and bushes unfold in May, and a homogenous dark green canopy has formed by mid-June. At this time the leaves of skunk cabbage begin to decay. They don't dry up and fall onto the ground to become part of the leaf litter that is slowly decomposed by fungi over the next year. Skunk cabbage has its own characteristic way of decaying. The leaves get small holes in them, begin to hang down, and parts turn black and somewhat slimy. Eventually the leaves sink to the ground and dissolve. This dissolution occurs rapidly, so that already by the end of July or early August the leaves are gone. You only find a few remnants of the bases of the leaf stalks. What dominated the appearance of the wetland in May has disappeared in August.

As strange as this way of decomposing at first seems, after studying the plant more intensively you begin to see how it fits with other characteristics. While growing, a plant is in its most fluid state. It then forms hard fibers, which, in biochemical terms, is a process of condensation and drying out. When the plant dies even more water is lost, and decay of the woody fibers sets in. Skunk cabbage stays in the watery phase; its substances don't condense and dry out. Therefore the dying leaves appear to evaporate, since they are mostly water, and almost no dry matter is left on the ground to decay. Skunk cabbage unfolds rapidly and disappears rapidly.

The Whole in the Part

Through this sketch I want to give you at least a partial view into the life of skunk cabbage. (For a more complete portrayal, see Holdrege 2000.) We can see its unique characteristics, but we can also see more. We can see how the various aspects of a plant's development, also in relation to its habitat, express certain unified tendencies.

When I see such relations, I sense that I'm finally beginning to actually meet and understand the plant, seeing through all the details to its unity and coherence. But at the same time, it's a new kind of territory. Where before I had seemingly solid objects—the different parts of the plant in their shape, size, consistency, etc.—now I'm dealing with the qualities that are expressed *through* these parts. And qualities aren't things.

Skunk cabbage expresses in many of its features a bud-like quality. Its flowers are housed in the large bud-like spathe, never extending out of this mantle. Skunk cabbage blooms in a bud at the time of year in which most flowers, later to unfold, are still tightly encased in their buds. Its flowers never reach the full light of day, and the parts of a flower that normally unfold are highly reduced. While the petals are missing altogether, the small, fleshy sepals, all tightly packed into a sphere, open only enough to let the stamens and style slightly protrude. The flower head remains a big, fleshy bud within the bud-like spathe.

When the plant grows, leaf upon leaf unwraps out of the large bud. Since the stem of the plant never elongates but remains in the ground, the leaves never grow apart. Instead, they form a funnel-shaped rosette. The rosette is only fully open, that is, the leaves spread out in horizontal fashion, when the leaves are dying. Their life is in the unfolding bud; being unfolded signals decay. And skunk cabbage never stops laying down new buds, so that an established plant harbors in its rootstock (rhizome) not only the spathe bud and leaf buds for the next season, which is typical for perennials, but also buds of each of a number of years to come.

We can go further and view these bud-like qualities in connection with skunk cabbage's dependency upon a wet environment. When I asked students in a field ecology course how they could determine where the wetland begins, they would often answer, "skunk cabbage shows you." Its roots need to be bathed in muddy soil throughout the year.

Skunk cabbage is not only dependent upon water, but also brings qualities of water—such as fluidity, movement, continuity, and the tendency to form surfaces—to expression. Early in spring, when stasis reigns in the wetland, skunk cabbage brings movement and life. The spathe grows out of the frozen ground and expresses in its form the congealed movement of spiraling surfaces. With the help of water, solid starch transforms into fluid sugar sap. Rising from the roots and rootstock, the fluid sugar is utilized in all growth processes. Moreover, large amounts of sugar are broken down to produce the warmth in the flower head. This transformation from solid starch to flowing sugar sap to radiating warmth is mediated by water and brings movement into the dormant landscape of early spring.

The radiating warmth in turn brings the air and insects into motion. When the leaves grow, you can almost see the water moving out of wet soil through the roots into the leaves, swelling and unfolding them. The leaves have a large, undulating surface that is like a conduit for water. They don't have a thick, waxy cuticle that prevents transpiration. As a result, water is continually flowing out of the soil, into and through the plant, and into the air, increasing the humidity of the lower layer of air in the wetland.

When skunk cabbage leaves decompose, they don't dry up and crumble; they dissolve. With few fibers, they consist mainly of water and air, as do the spathe and flowers, and disintegrate into these elements. Skunk cabbage embodies wateriness, growing and decaying in its watery world.

The Unity of the Organism

As the process of knowing unfolds—the conversation with the plant—you begin to see the unity of the plant. The remarkable thing is that when you build exact pictures over and over, moving from one characteristic to the next, patterns emerge. You begin to recognize how the characteristics express a whole—the unity begins to reveal itself. When you go back to characteristics you have studied before, they may suddenly express the unity you have discovered through another part. You have an "aha" experience in which you recognize connections between what previously appeared to be separate facts. You see a common watery, bud-like quality in the form and consistency of spathe, flower head, and leaves. Skunk cabbage reveals the fluid quality of water in the way it unfolds and decays, as well as in its undulating, flowing forms. And in all of these characteristics you can see a vivid picture of early spring—a plant that is bud-like in so many ways and yet unfolds to bring the first life, warmth, and movement to a still slumbering habitat.

While you have to work hard to get to such insights, you cannot force them. If you try to, you can be pretty sure they won't come. This is a stage of knowing where you have to learn patience. You prepare the ground, but the moment of seeing always involves an act of grace. Or maybe we could just say: we have to wait until the world speaks. As Goethe describes it:

> I persist until I have discovered a pregnant point from which much may be derived, or rather—since I am careful in my work

> and observations—one which yields several things, offering them up of its own accord. If some phenomenon appears in my research, and I can find no source for it, I let it stand as a problem. This approach has proven quite advantageous over the years. When I found I could not solve the riddle of the origin and context of some problem, I had to let it lie for a long time; but at some moment, years later, enlightenment came in the most wonderful way. (Goethe 1995, 41, translation modified by Craig Holdrege)

Once you've come to understand a plant in this way, you never encounter it with the remark, "Oh, that's just a skunk cabbage." Rather, you meet it with expectation and interest, wondering what else it has to show you. And this attitude begins to inform your overall orientation toward nature. Any other plant, beetle, or bird you see appears immediately as a riddle and not a thing. You know that each carries within itself—as you've experienced in skunk cabbage—a whole, unique world that's just waiting to be disclosed.

This is a key result of working in a Goethean way. You connect with the world. You are no longer an onlooker. And you have met something that now *informs* you. You have had a productive conversation. Now you also know from the inside out why it would be a poignant loss if, through unconsidered human action (such as the draining of woodland swamps for housing developments and agriculture), skunk cabbage were to disappear from the world. A unique quality—a special voice—would be missing from nature's chorus. You cannot just stand by and simply let such a thing happen. Of course there is no prescription for human action and for what any given individual will do. But at least the world is no longer an impersonal "out there."

One of the problems with talking *about* practicing the Goethean approach to science is that the essence is in the doing itself. That's why the description of concrete examples is so important. I'll conclude this chapter by pulling back from the example and presenting some of the key features of a Goethean approach as I've described it.

Preparing the Ground—A New Attitude of Mind

All science has its roots in human questioning and the search for understanding. As far as I can see, most people who are drawn to Goethe's approach to science recognize in it a way of understanding nature that can take them beyond the boundaries of what has developed as mainstream science. At the heart of the Goethean approach is the realization that as a scientist I must develop new capacities in order to do justice to nature in my work. It's not just a matter of developing new instruments or refining the intellect, but developing new ways of knowing that can illuminate the phenomena in ways that science has largely neglected (or even deemed unscientific).

Out of this awareness arises the striving to develop a gentle sensibility that does not violate the phenomena in the process of getting to know them. It's an active conversation, but one in which I hope the other—as something in its own right—can reveal itself. As Goethe writes, the scientist strives to "find the measure for what he learns, the data for judgment, not in himself but in the sphere of what he observes" (Goethe 1995, 11, written in 1792). This is the attitude that Goethe suggests with his expression "delicate empiricism" and that I've described above through the metaphor of conversation. As a kind of underlying intentionality, it permeates all the work one does and grows as a capacity the more one works.

For this attitude of mind to actually inform every aspect of one's work means removing many obstacles—habits of mind that have us search for single causes, for general theories, for reductive explanations. In the end it means, in the words of Owen Barfield, ridding oneself of all "residues of unresolved positivism" (in Sugarman 1976, 13–15). This is not an easy task, and one that never ends. Yet the striving (and some success!) is absolutely necessary if nature is to show herself from new sides. In his description of phenomenology as a new way of viewing, Edmund Husserl spoke out of the soul of the Goethean scientist:

> That we should set aside all previous habits of thought, see through and break down the mental barriers which these habits have set along the horizons of our thinking, and in full intellectual freedom proceed to lay hold on those genuine philosophical problems still awaiting completely fresh formulation which the liberated horizons on all sides disclose to us—these are hard demands. Yet nothing less is required. . . . A new way of looking at things is necessary, one that contrasts at every point with the natural attitude of experience and thought. To move freely along this new way without ever reverting to the old viewpoints,

> to learn to see what stands before our eyes, to distinguish, to describe, calls, moreover, for exacting and laborious studies. (Husserl 1962, 39)

Practicing the Goethean Approach to Science

The Riddle

This is the beginning of any investigation. I am drawn to a particular phenomenon and want to get to know it better. I've met something in the world that is a riddle I want to attend to. And because each person has a different biography—carries a unique world within herself—and is drawn to different features of the world, there is an endless and beautiful array of possible questions and areas of focus. I have colleagues who are physicists, chemists, ecologists, botanists, and zoologists. They are not only investigating different realms of phenomena, but they also take somewhat different approaches based on who each of them is. This does not make the work "subjective," but merely points to the fact that in any scientific endeavor the subject as a particular being is actively at work. And the riddle that draws a particular person is the beginning of a pathway into the world that is specific, but can be shared with others. (We live, after all, in one world.)

Into the Phenomena

This is exploration, getting to know the phenomena. As Goethe wrote in connection with his work in optics: "The greatest accomplishments come from those who never tire in exploring and working out every possible aspect and modification of every bit of empirical evidence, every experience" (Goethe 1995, 15, written in 1792).

You really have to get to know the phenomena you're dealing with from as many sides as possible. If you're doing experiments, then it's a matter of varying them in a methodical way to build up a rich picture. It's not about proving (or falsifying) a particular hypothesis (Ribe and Steinle 2002). In studying a living organism, you want to gain a many-sided picture of the life of the organism and its relation to its environment. In this work you make your own observations, but you also interact with and utilize the work of others (which typically requires disentangling theory and interpretation). Here is where a research community evolves. As Goethe writes, "What applies in so many other human enterprises is also true here [in science]: the interest of many focused on a simple point can produce excellent results. . . . I have always found working together with others so advantageous that I have every reason to continue doing so" (Goethe 1995, 12–13, translation modified by Craig Holdrege).

Since the phenomena are endless, this work is also without end. I can never get "all the facts," but my goal is also not an encyclopedic totality of information. It's more that I never cease to be interested in what the phenomena—perhaps some unassuming, seemingly esoteric detail—may reveal to me about the world. In my own work I often find that we don't know nearly enough about the animal or plant I'm studying. I do extensive literature searches and speak with experts, am enriched by all I find, but am usually left feeling I'd love to know much more. I also discover how theory-burdened so much of science is, with a small number of facts being marshaled to apparently support grand ideas.

Exact Picture Building

While getting to know the phenomena, I intensify my experience through exact picture building—Goethe's exact sensorial imagination. At first this may be a completely separate activity from being out and observing. I retreat from observation and quietly build up a precise inner picture of what I've experienced. The more I've done this, the more I find that my observing and perceiving become dynamic and full of life. I become active while perceiving, following inwardly the shapes, colors, smells, or tones as I observe. I sculpt the shapes while looking. This is where I notice how the picture-building as an exercise becomes integrated into my concrete interaction with the phenomena. I begin to see more intensely.

This work helps me to enter more deeply into the phenomenal world. It also gives my experience of the organism more continuity. The connectedness of all the details within the organism itself also becomes a connectedness within me.

I have come to see this activity of exact sensorial imagination to be the counter pole to theory or model building in traditional science. In both cases we are inwardly active. But in exact sensorial imagination, the work of concrete picturing—building images and letting the one transform into the other—keeps us close to the phenomena. We close the gaps that are given through our discrete observations and in this sense go beyond what perception gives us, but our whole intention is to take in the world. We activate the imagination to get as close to the phenomenal reality as possible. By contrast, in theory building we assemble thoughts out of ourselves that are distinct from the phenomena. These thoughts provide a conceptual framework within which we view the phenomena. We may even develop a mental model, such as the lock and key model of enzyme action I learned in college. Such models are not developed out of an interaction with the phenomena, but abstracted from a wholly different realm of experience (opening and locking doors) to help us explain what may be going on "behind the scenes" at a molecular level. Of course, any thoughtful scientist will admit that such models are just a crutch and ultimately only a kind of reified stand-in for more abstract conceptual relations.

Because in model and theory building we step back from the phenomena and construct a coherent conceptual edifice, we may begin seeing everywhere embodiments of our models and theories rather than the things themselves. We see a moving survival strategy rather than a fleeing prairie dog; we see a genetic defect rather than a person. Theories tend to take on a life of their own. This is what Alfred North Whitehead called the fallacy of misplaced concreteness (1967, especially chapters 3 and 4). Goethe was highly sensitive to our human propensity to substitute abstractions for reports of experience. He spoke of the continual danger of a concept becoming a despotic "lethal generality" (1995, 61). In contrast, exact sensorial imagination has the cathartic effect of reorienting our attention to the phenomena, while dissolving hard-and-fast ideas through mental molding and remolding.

Seeing the Whole

This is the "step" that we've been preparing for in all the other work. Or, stated more accurately, this is what can reveal itself in the course of one's striving to get to know the phenomena. As I said above, it is an experience of seeing unifying relations, which may or may not happen during any investigation. When it occurs, it fills you with the greatest joy and you realize: "Now I am knowing." We can use the word intuition here as long as we don't think of something vague, but rather a nondiscursive form of seeing connections that is comparable to the experience one can have most purely in mathematical insight.

In the example of skunk cabbage I tried to show how you can see a bud-like, watery quality in various characteristics of the plant. Its wholeness speaks through its parts and its relation to the environment. If you imagine this mode of cognition applied on a larger scale, you come to what Goethe writes about as the "archetypal phenomena" in his color work, or the "type" (*Typus*) and the archetypal animal or plant (*Urpflanze; Urtier*) in his biological studies. (He also speaks of "entelechy," or "idea.") What term one uses is much less important than the quality of knowing itself. Here's how he describes the whole process, brilliantly condensed into a few sentences, that leads to a seeing that goes beyond, but is fully rooted in, empirical observation:

> If I look at the created object, inquire into its creation, and follow this process back as far as I can, I will find a series of steps. Since these are not actually seen together before me, I must visualize them in my memory so that they form a certain ideal whole. At first I will tend to think in terms of steps, but nature leaves no gaps, and thus, in the end, I will have to see this progression of uninterrupted activity as a whole. I can do so by dissolving the particular without destroying the impression. . . . If we imagine the outcome of these attempts, we will see that empirical observation finally ceases, inner beholding of what develops begins, and, at last, the idea can be brought to expression. (Goethe, 1995, 75, translation modified by Craig Holdrege)

If you don't pay attention to the process and context out of which Goethe speaks about bringing an idea to expression, you could imagine "idea" to be something abstract or bloodless ("just another theory"). But it's not. It has much more the nature of seeing a being. That's why Goethe was so distraught when Friedrich Schiller reacted to his description of the archetypal plant by stating, "That is not an observation from experience. It is an idea." Goethe responded: "Then I may rejoice that I have ideas without knowing it, and can even see them with my own eyes" (Goethe 1995, 18–21).

So when Goethe says there is a "delicate empiricism which makes itself utterly identical with the object, thereby becoming true theory" (1995, 307), then "theory" is to be understood in the sense of the ancient Greeks as a "seeing of the mind" or "beholding" and not as the abstract "theory" we know from modern science. If we use the term "idea," then we must think of an idea that Goethe could, in the end, see sensibly/supersensibly in every plant. Reflecting on his botanical studies, Goethe writes in 1831, near the end of his life:

> A challenge . . . hovered in my mind at that time [1787] in the sensuous form of a supersensuous plant archetype [*Urpflanze*]. I traced the variations of all the forms as I came upon them. In Sicily, the final goal of my [Italian] journey, the conception of the original identity of all plant parts had become completely clear to me; and everywhere I attempted to pursue this identity and to catch sight of it again. . . . Only a person who has himself experienced the impact of a fertile idea . . . will understand what passionate activity is stirred in our minds, what enthusiasm we feel, when we glimpse in advance and in its totality something which is later to emerge in greater and greater detail in the manner suggested by its early development. Thus the reader must surely agree that, having been captured and driven by such an idea, I was bound to be occupied with it, if not exclusively, nevertheless during the rest of my life. (Goethe 1989, 162)

So finding the fertile idea is at once a completion of a process and the beginning of a new one. As an end, it brings us full circle to a more conscious glimpse of the being—the riddle—that formed the starting point of the investigation. As a beginning, it is the soil for further work and vital new insights. Goethe's approach to science is itself a fertile idea that still has ample life to unfold.

Acknowledgments

This book grew out of our work at The Nature Institute, a small nonprofit organization financially supported through the generosity of individual donors and foundation grants. We would like to thank all our contributing "Friends of The Nature Institute" and also thank and recognize those organizations that, over the past eight years, have supported our research and writing related to genetics and genetic engineering: the Education Foundation of America, Evidenzgesellschaft, Foundation for Rudolf Steiner Books, GTS Treuhand, Future Value Fund, Mahle-Stiftung, New Earth Foundation, Rudolf Steiner Charitable Trust, RSF Shared Gifting Group, Rudolf Steiner-Fonds für Wissenschaftliche Forschung, Software AG Stiftung, T. Backer Fund, Waldorf Schools Fund, and the Waldorf Educational Foundation.

The chapters in this book are based on essays we have written over the course of the past ten years. Below we credit the original essays, most of which have been substantially revised and updated for this book. The articles published in *NetFuture* can be found at netfuture.org, and the articles published in *In Context* are available at natureinstitute.org/pub/ic.

Chapter 1, "Sowing Technology," by Craig Holdrege and Steve Talbott, appeared in *Sierra* (July/August 2001): 34–39, 72. A version closer to the one in this volume was published in *NetFuture* 123 (October 9, 2001).

Chapter 2, "Golden Genes," by Craig Holdrege and Steve Talbott, appeared in *NetFuture* 108 (July 6, 2000).

Chapter 3, "Will Biotech Feed the World? The Broader Context," by Craig Holdrege, was published in 2005 on The Nature Institute's Web site: http://natureinstitute.org/txt/ch/feed_the_world.htm.

Chapter 4, "Should Genetically Modified Foods Be Labeled?" by Craig Holdrege, appeared in *NetFuture* 135 (August 29, 2002).

Chapter 5, "Genes Are Not Immune to Context: Examples from Bacteria," by Craig Holdrege, appeared in *In Context* 12 (fall 2004): 11–12.

Chapter 6, "The Gene: A Needed Revolution?" by Craig Holdrege, appeared in *In Context* 14 (fall 2005): 14–17.

Chapter 7, "Life beyond Genes: Reflections on the Human Genome Project," by Craig Holdrege and Johannes Wirz, appeared in *In Context* 5 (spring 2001): 14–19.

Chapter 8, "Me and My Double Helixes," by Steve Talbott, appeared in *NetFuture* 144 (April 29, 2003).

Chapter 9, "Logic, DNA, and Poetry," by Steve Talbott, appeared in *NetFuture* 160 (January 25, 2005), and also in *The New Atlantis* 8 (spring 2005).

Chapter 10, "The Cow: Organism or Bioreactor?" by Craig Holdrege, is drawn from the introduction and chapter 5 of *Genetics and the Manipulation of Life,* by Craig Holdrege (Great Barrington, Mass.: Lindisfarne Press, 1996), and from "The Cow: Organism or Bioreactor?" *Orion* (winter 1997): 28–32.

Chapter 11, "The Forbidden Question," by Craig Holdrege and Steve Talbott, appeared in very different versions in *Orion* (July/August 2006): 24–31, and *NetFuture* 166 (January 16, 2007), as "Science's Forbidden Question."

Chapter 12, "What Does It Mean to Be a Sloth?" by Craig Holdrege, appeared in the *Newsletter of the Society for the Evolution of Science* 14, no. 1 (winter 1998): 1–26, as "The Sloth: A Study in Wholeness," and in *NetFuture* 97 (November 3, 1999).

Chapter 13, "The Language of Nature," by Steve Talbott, appeared in *The New Atlantis* 15 (winter 2007): 41–76, and in *NetFuture* 167 (March 15, 2007), *NetFuture* 168 (April 13, 2007), and *NetFuture* 169 (May 10, 2007).

Chapter 14, "Doing Goethean Science," by Craig Holdrege, appeared in *Janus Head* 8, no. 1 (winter 2005): 27–52.

References

Aiello, Annette. 1985. "Sloth Hair: Unanswered Questions." In *The Evolution and Ecology of Armadillos, Sloths, and Vermilinguas,* ed. G. Gene Montgomery, 213–18. Washington, D.C.: Smithsonian Institution Press.

Alliance for Biointegrity. n.d. "Key FDA Documents Revealing (1) Hazards of Genetically Engineering Foods and (2) Flaws with How the Agency Made Its Policy." Available online: http://www.biointegrity.org/list.html.

Altieri, Miguel A. 2000. "Biotech Will Not Feed the World." *San Francisco Chronicle,* March 30, A-27. Available online: http://sfgate.com/cgi-bin/article.cgi?f=/c/a/2000/03/30/ED38656.DTL.

Amrine, Frederick. 1987. "Goethe and the Sciences: An Annotated Bibliography." In *Goethe and the Sciences: A Reappraisal,* ed. Frederick Amrine et al., 389–487. Boston: D. Reidel Publishing Company.

———. 1998. "The Metamorphosis of the Scientist." In Seamon and Zajonc 1998, 33–54.

Andrews, Lori. 2002. "Genes and Patent Policy: Rethinking Intellectual Property Rights." *Nature Reviews Genetics* 3:803–8.

Angier, Natalie. 2000. "A Pearl and a Hodgepodge: Human DNA." *New York Times,* June 27, A1 and A21.

Anson, B. J. 1950. *Atlas of Human Anatomy.* Philadelphia: W. B. Saunders.

Barfield, Owen. 1965. *Saving the Appearances.* 1957. New York: Harcourt, Brace, and World.

———. 1967. *Speaker's Meaning.* Middletown, Conn.: Wesleyan Univ. Press.

———. 1971. *What Coleridge Thought.* Middletown, Conn.: Wesleyan Univ. Press.

———. 1973. *Poetic Diction: A Study in Meaning.* 1928. Middletown, Conn.: Wesleyan Univ. Press.

———. 1977. "The Rediscovery of Meaning." In *The Rediscovery of Meaning and Other Essays.* Middletown, Conn.: Wesleyan Univ. Press.

Beaber, J., B. Hochhut, and B. Waldor. 2004. "SOS Response Promotes Horizontal Dissemination of Antibiotic Resistance Genes." *Nature* 427:72–74.

Beck, Martha. 1999. *Expecting Adam.* New York: Random House.

Beebe, William. 1926. "The Three-Toed Sloth." *Zoologica* 7:1–67.

Benbrook, Charles M. 2001. Personal communication with the authors.

———. 2002. "Economic and Environmental Impacts of First Generation Genetically Modified Crops: Lessons from the United States." *International Institute for Sustainable Development—Trade Knowledge Network.* Available online: http://www.tradeknowledgenetwork.net/pdf/tkn_firstgen_gmo_us.pdf.

———. 2004. "Genetically Engineered Crops and Pesticide Use in the United States: The First Nine Years." *BioTech InfoNet Technical Paper Number 7.* Available online: http://www.biotech-info.net/highlights.html#technical_papers.

Benzer, Seymour. 1962. "The Fine Structure of the Gene." *Scientific American* (January) (reprint): 2.

Beurton, Peter J., Raphael Falk, and Hans-Jörg Rheinsberger, eds. 2000. "Introduction." In *The Concept of the Gene in Development and Evolution*, ix–xiv. Cambridge: Cambridge Univ. Press.

Bialy, H. 1991. "Transgenic Pharming Comes of Age." *Bio/Technology* 9:786–88.

Bjedov, I., O. Tenaillon, B. Gérard, et al. 2003. "Stress-Induced Mutagenesis in Bacteria." *Science* 300:1404–9.

Björkman, J., I. Nagaev, O. G. Berg, et al. 2000. "Effects of Environment on Compensatory Mutations to Ameliorate Costs of Antibiotic Resistance." *Science* 287:1479–82.

Bohm, David. 1971. *Causality and Chance in Modern Physics.* Philadelphia, Pa.: Univ. of Pennsylvania Press.

———. 1996. *On Dialogue.* New York: Routledge Classics.

Bohner, Horst. 2003. "What about Yield Drag on Roundup Ready Soybean?" *Ontario Ministry of Agriculture, Food, and Rural Affairs.* Available online: http://www.omafra.gov.on.ca/english/crops/field/news/croptalk/2003/ct_0303a9.htm.

Bortoft, Henri. 1996. *The Wholeness of Nature: Goethe's Way toward a Science of Conscious Participation in Nature.* Hudson, N.Y.: Lindisfarne Press.

Bourlière, François. 1964. *The Natural History of Mammals.* New York: Alfred A. Knopf. 252–57.

Brady, Ronald H. 2002. "Perception: Connections between Art and Science." Available online: http://natureinstitute.org/txt/rb/art/perception.htm.

Britton, W. S. 1941. "Form and Function in the Sloth." *Quarterly Review of Biology* 16:13–43, 190–207.

Bruecher, H. 1982. *Die Sieben Säulen der Welternährung.* Frankfurt am Main: Waldemar Kramer.

Bull, James, and Bruce Levin. 2000. "Mice Are Not Furry Petri Dishes." *Science* 287:1409–10.

Burian, R. M. 1985. "On Conceptual Change in Biology: The Case of the Gene." In *Evolution at a Crossroads: The New Biology and the New Philosophy of*

Science, ed. D. J. Depew and B. H. Weber, 21–42. Cambridge, Mass.: MIT Press.

Cabib S., C. Orsini, M. Le Moal, et al. 2000. "Abolition and Reversal of Strain Differences in Behavioral Responses to Drugs of Abuse after a Brief Experience." *Science* 289:463–65.

Cahill J. F., J. P. Castelli, and B. B. Casper. 2001. "The Herbivory Uncertainty Principle: Visiting Plants Can Alter Herbivory." *Ecology* 82:307–12.

Cellini, F., A. Chesson, I. Colquhoun, et al. 2004. "Unintended Effects and Their Detection in Genetically Modified Crops." *Food and Chemical Toxicology* 42:1089–125.

Claverie, J-M. 2001. "What If There Are Only 30,000 Human Genes?" *Science* 291:1255–57.

Code of Federal Regulations. n.d. *Code of Federal Regulations,* Title 21, Chapter 1, especially parts 101 and 102. Available online: http://www.access.gpo.gov/cgi-bin/cfrassemble.cgi?title=200121.

Conway, Gordon, and Susan Sechler. 2000. "Helping Africa Feed Itself." *Science* 289:1685.

Coomaraswamy, Ananda K. 1977. *Coomaraswamy: 1: Selected Papers—Metaphysics.* Ed. Roger Lipsey. Princeton: Princeton Univ. Press.

Cornford, F. M. 1957. *From Religion to Philosophy: A Study in the Origins of Western Speculation.* New York: Harper and Brothers.

Daniell, Henry. 1997. "Transformation and Foreign Gene Expression in Plants Mediated by Microprojectile Bombardment." In *Recombinant Gene Expression Protocols,* ed. Henry Daniell, 463–90. Totowa, N.J.: Humana Press.

Darnell, J., H. Lodish, and D. Baltimore. 1986. *Molecular Cell Biology.* New York: Scientific American Books.

Dayuan, Xue. 2002. "The Environmental Impact of BT Cotton in China." Available online: http://archive.greenpeace.org/geneng/reports/env_impact_eng.pdf.

Dennett, Daniel C. 1995. *Darwin's Dangerous Idea: Evolution and the Meanings of Life.* New York: Simon and Schuster.

Dijksterhuis, E. J. 1961. *The Mechanization of the World Picture.* Oxford: Oxford Univ. Press.

Dohoo, I. R., K. Leslie, L. DesCôteaux, et al. 2003a. "A Meta-Analysis Review of the Effects of Recombinant Bovine Somatotropin, 1: Methodology and Effects on Production." *Canadian Journal of Veterinary Research* 67:241–51.

Dohoo, I. R., L. DesCôteaux, K. Leslie, et al. 2003b. "A Meta-Analysis Review of the Effects of Recombinant Bovine Somatotropin, 2: Effects on Animal Health, Reproductive Performance, and Culling." *Canadian Journal of Veterinary Research* 67:252–64.

Duboule, D., and A. Wilkins. 1998. "The Evolution of 'Bricolage.'" *Trends in Genetics* 14:54–59.

Duffy, Michael. 2001. "Who Benefits from Biotechnology?" Talk at the American Seed Trade Association, December 2001. Available online: http://www.newfarm.org/depts/gleanings/1203/duffybiotech_print.shtml.

Eddington, Sir Arthur. 1920. *Space, Time, and Gravitation.* Cambridge: Cambridge Univ. Press.

Edelglass, S., G. Maier, H. Gebert, and J. Davy. 1997. *The Marriage of Sense and Thought.* Great Barrington, Mass.: Lindisfarne Press.

Einstein, Albert. 1954. *Ideas and Opinions.* Trans. Sonja Bargmann. New York: Crown Publishers.

Ellul, J. 1990. *The Technological Bluff.* Trans. Geoffrey W. Bromiley. Grand Rapids, Mich.: Eerdmans.

Erdman, J., T. Beirer, and E. Gugger. 1993. "Absorption and Transport of Carotenoids." In *Carotenoids in Human Health,* ed. L. Canfield, N. Krinsky, and J. Olson, 76–85. New York: New York Academy of Sciences.

FDA. 1995. "FDA's Policy for Foods Developed by Biotechnology." CFSAN Handout. Available online: http://www.cfsan.fda.gov/~lrd/biopolcy.html.

———. 2004. *An FDA Overview: Protecting Consumers, Protecting Public Health,* produced by FDA Office of Public Affairs. Available online: http://www.fda.gov/oc/opacom/fda101/fda101text.html.

Federal Register. 1992. *Federal Register* 54 (104): 22991.

———. 1994. *Federal Register* 59 (98): 26700–711.

Feldbaum, Carl B. 1998. "Can Bioengineers Feed the Planet?" *New York Times,* letter to the editor, July 23, A24.

Fernandez-Cornejo, J., and W. D. McBride. 2002. *The Adoption of Bioengineered Crops.* U.S. Department of Agriculture, Economic Research Service, Agricultural Economic Report No. 810.

Feynman, Richard. 1967. *The Character of Physical Law.* Cambridge, Mass.: MIT Press.

Feynman, Richard P., Robert B. Leighton, and Matthew Sands. 1963. *The Feynman Lectures on Physics,* vol. 1. Reading, Mass.: Addison-Wesley.

Food and Agriculture Organization of the United Nations. 2002. *World Agriculture: Towards 2015/2030 Summary Report.* Rome. Available online: http://www.fao.org/docrep/004/y3557e/y3557e00.htm.

———. 2004. *The State of Food and Agriculture: 2003–2004.* Rome. Available online: http://www.fao.org/docrep/006/y5160e/y5160e00.htm.

———. 2005. *The State of Food Insecurity in the World: 2005.* Rome. Available online: http://www.fao.org/docrep/008/a0200e/a0200e00.htm.

Foster, Patricia. 2004. "Adaptive Mutation in *Escherichia coli.*" *Journal of Bacteriology* 186:4846–52.

Fray, R., A. Wallace, P. Fraser, et al. 1995. "Constitutive Expression of a Fruit Phytoene Synthase Gene in Transgenic Tomatoes Causes Dwarfism by Redirecting Metabolites from Gibberellin Pathway." *Plant Journal* 8:693–701.

Freese, William, and David Schubert. 2004. "Safety Testing and Regulation of Genetically Engineered Foods." *Biotechnology and Genetic Engineering Reviews* 21:299–324.

Fresco, Louise. 2003. "Which Road Do We Take? Harnessing Genetic Resources and Making Use of Life Sciences, a New Contract for Sustainable Agriculture." Talk given at an EU Discussion Forum, Brussels, January 30–31, 2003. Available online: http://www.fao.org/ag/magazine/fao-gr.pdf.

Galilei, Galileo. 1957. *Discoveries and Opinions of Galileo.* Trans. and with an introduction and notes by Stillman Drake. New York: Random House.

Gardner, G., and B. Halweil. 2000. *Underfed and Overfed: The Global Epidemic of Malnutrition,* Worldwatch Paper 150. Washington, D.C.: Worldwatch Institute.

Gehring, W., and K. Ikeo. 1999. "Pax 6: Mastering Eye Morphogenesis and Eye Evolution." *Trends in Genetics* 15:371–77.

Gelbart, William. 1998. "Databases in Genomic Research." *Science* 282:659–61.

Gertz, J. M., W. K. Vencill, and N. S. Hill. 1999. "Tolerance of Transgenic Soybean (Glycine max) to Heat Stress." *The Brighton Conference: Weeds.* Surrey, U.K.: The Council, 835–40.

Gilbert, W. 1991. "Towards a Paradigm Shift in Biology." *Nature* 349:99.

———. 1992. "A Vision of the Grail." In *The Code of Codes: Scientific and Social Issues in the Human Genome Project,* ed. D. J. Kevles and L. Hood, 83–97. Cambridge, Mass.: Harvard Univ. Press.

Gin, M. 1975. *Ricecraft.* San Francisco: Yerba Buena Press.

Glickman, Dan. 1997. Remarks to International Grains Council, London, England (June 19). Available online: http://www.usda.gov/news/releases/1997/06/0196.

Goethe, Johann W. von. 1982. *Italian Journey.* San Francisco: North Point Press.

———. 1989. *Goethe's Botanical Writings.* Woodbridge, Conn.: Ox Bow Press.

———. 1995. *Goethe: Scientific Studies*, vol. 12 (Collected Works, ed. Douglas Miller). Princeton: Princeton Univ. Press.

Goffart, M. 1971. *Form and Function in the Sloth.* Oxford and New York: Pergamon Press.

Goldstein, Kurt. 1963. *The Organism.* Boston: Beacon Press. (Originally published in 1939. Reprinted in 1995 by Zone Books in New York.)

Gordon, Jon W. 1999. "Genetic Enhancement in Humans." *Science* 283 (March 26): 2023–24.

Griffiths, Paul E., and Eva M. Neumann-Held. 1999. "The Many Faces of the Gene." *BioScience* 49:656–62.

Groh, T., and S. McFadden. 1997. *Farms of Tomorrow.* Junction City, Oreg.: Biodynamic Farming and Gardening Association.

Grzimek, Bernhard. 1975. *Grzimek's Animal Life Encyclopedia,* vol. 11 (Mammals II). New York: Van Nostrand Reinhold.

Guerinot, M. 2000. "The Green Revolution Strikes Gold." *Science* 287:241–43.

Gunning R. V., H. T. Dang, F. C. Kemp, et al. 2005. "New Resistance Mechanism in *Helicoverpa armigera* Threatens Transgenic Crops Expressing *Bacillus thuringiensis* Cry 1Ac Toxin." *Applied and Environmental Microbiology* 71:2558–63.

Hadden, S. 2000. "How Much Use Is the Human Genome Project?" *Nature* 406:541–42.

Hadley, G. L., C. A. Wolf, and S. B. Harsh. 2006. "Dairy Culling Patterns, Explanations, and Implications." *Journal of Dairy Science* 89:2286–96.

Halder, G., P. Callaerts, and W. Gehring. 1995. "Induction of Ectopic Eyes by Targeted Expression of the Eyeless Gene in *Drosophila*." *Science* 267:1788–92.

Hallman, W., W. Hebden, C. Cute, et al. 2004. *Americans and GM Food: Knowledge, Opinion, and Interest in 2004.* FPI publication number RR-1104-007. Rutgers, N.J.: Food Policy Institute, Cook College, Rutgers (http://www.foodpolicyinstitute.org).

Hardell, Lennart, and Mikael Eriksson. 1999. "A Case-Control Study of Non-Hodgkin Lymphoma and Exposure to Pesticides." *Cancer* 85 (March 15): 1353–60.

Hastings, P. J., H. J. Bull, J. R. Klump, and S. M. Rosenberg. 2000. "Adaptive Amplification: An Inducible Chromosomal Instability Mechanism." *Cell* 103:723–31.

Henderson, E. 1999. *Sharing the Harvest: A Guide to Community-Supported Agriculture.* White River Junction, Vt.: Chelsea Green.

Holdrege, Craig. 1996. *Genetics and the Manipulation of Life: The Forgotten Factor of Context.* Hudson, N.Y.: Lindisfarne Press.

———. 1998. "Seeing the Animal Whole: The Example of Horse and Lion." In Seamon and Zajonc 1998, 213–32.

———. 2000. "Skunk Cabbage (*Symplocarpus foetidus*)." *In Context* 4:12–18. Available online: http://natureinstitute.org/pub/ic/ic4/skunkcabbage.htm.

———. 2004a. *The Flexible Giant: Seeing the Elephant Whole.* Ghent, N.Y.: The Nature Institute.

———. 2004b. "Science Evolving: The Case of the Peppered Moth." In *Writing the Future: Progress and Evolution,* ed. David Rothenburg and Wandee Pryor. Cambridge: MIT Press. (An earlier version of this paper is available online: http://natureinstitute.org/txt/ch/moth.htm.)

———. 2005a. "The Forming Tree." *In Context* 14:18–23.

———. 2005b. *The Giraffe's Long Neck: From Evolutionary Fable to Whole Organism.* Ghent, N.Y.: The Nature Institute.

———. 2006. "Can We See with Fresh Eyes?" *In Context* 16:18–23.

Horowitz, Norman H. 1956. "The Gene." *Scientific American* (October) (reprint): 4.

Husserl, Edmund. 1962. *Ideas: General Introduction to Pure Phenomenology.* London: Collier Books.

ICIPE. 2002/2003. *ICIPE Annual Scientific Report: Habitat Management Strategies for Control of Stemborers and Striga Weed in Cereal-Based Farming Systems in Eastern Africa.* Available online: http://www.push-pull.net/PDF%20files/01-Push-pull.pdf.

IDRC/CRDI. 2004. *Fixing Health Systems.* Available online: http://www.idrc.ca/tehip.

International Human Genome Sequencing Consortium. 2001. "Initial Sequencing and Analysis of the Human Genome." *Nature* 409:860–921.

———. 2004. "Finishing the Euchromatic Sequence of the Human Genome." *Nature* 431:931–45.

Jackson, M. 1995. "Protecting the Heritage of Rice Biodiversity." *GeoJournal* 35:267–74.

Jordan, Carl. 2002. "Genetic Engineering, the Farm Crisis, and World Hunger." *BioScience* 52:523–29.

Juskevich, J., and C. G. Guyer. 1990. "Bovine Growth Hormone: Human Food Safety Evaluation." *Science* 249:875–84.

Kangmin, Li. 1998. "Rice Aquaculture Systems in China: A Case of Rice-Fish Farming from Protein Crops to Cash Crops." *Proceedings of the Internet Conference on Integrated Bio-Systems.* Available online: http://www.ias.unu.edu/proceedings/icibs/li/paper.htm.

Kaplan, Warren. 2001. "Biotech Patenting 101." *GeneWatch* 14 (3): 6–9.

Kay, Lily E. 2000. *Who Wrote the Book of Life? A History of the Genetic Code.* Stanford, Calif.: Stanford Univ. Press.

Kell, D. 1991. "Lacuna Seeker." *Nature* 350:268.

Keller, David R., and E. Charles Brummer. 2002. "Putting Food Production in Context: Toward a Postmechanistic Agricultural Ethic." *BioScience* 52:264–71.

Keller, Evelyn Fox. 2000. *The Century of the Gene.* Cambridge, Mass.: Harvard Univ. Press.

Khan, Z., W. Overholt, and A. Hassana. n.d. "Utilization of Agricultural Biodiversity for Management of Cereal Stemborers and Striga Weed in Maize-Based Cropping Systems in Africa—A Case Study." *International Centre of Insect Physiology and Ecology.* Available online: http://www.push-pull.net/PDF%20files/cs-agr-cereal-stemborers.pdf.

Kitcher, P. 1992. "Gene: Current Usages." In *Keywords in Evolutionary Biology,* ed. E. F. Keller and E. A. Lloyd, 128–31. Cambridge, Mass.: Harvard Univ. Press.

Knutson, Roger M. 1974. "Heat Production and Temperature Regulation in Eastern Skunk Cabbage." *Science* 186:746–47.

Koechlin, F. 2000a. "Pro-Vitamin A-Reis—Die Grosse Illusion?" Available online: http://www.blauen-institut.ch/Pg/pT/a_T.html.

———. 2000b. "Natural Success Stories—The ICIPE in Kenya." Available online: http://www.blauen-institut.ch/Tx/tF/tfNaturalSuccess.html.

Kolata, Gina. 1993. "Unlocking the Secrets of the Human Genome." *New York Times,* November 30, C1 and C8.

Kranich, Ernst-Michael. 1995. *Wesensbilder der Tiere.* Stuttgart: Verlag Freies Geistesleben.

———. 1999. *Thinking beyond Darwin.* Great Barrington, Mass.: Lindisfarne Press.

Krimsky, Sheldon. 2003. *Science in the Private Interest.* Lanham, Md.: Rowman and Littlefield.

Kshirsagar, K. G., and S. Pandey. 1997. "Diversity of Rice Cultivars in a Rainfed Village in the Orissa State of India." *International Development Research Centre.* Ottawa, Canada. Available online: http://www.idrc.ca/en/ev-85117-201-1-DO_TOPIC.html.

Lappé, Francis Moore, et al. 1998. *World Hunger: Twelve Myths.* New York: Grove Press.

Lauda, S. M., R. W. Pemberton, M. T. Johnson, and P. A. Follet. 2003. "Nontarget Effects—The Achilles' Heel of Biological Control?" *Annual Review of Entomology* 48:365–96.

Lewis, C. S. 1965. *The Abolition of Man.* New York: Macmillan.

Lindgren, L., K. Stålberg, and A. Höglund. 2003. "Seed-Specific Overexpression of an Endogenous Arabidopsis Phytoene Synthase Gene Results in Delayed Germination and Increased Levels of Carotenoids, Chlorophyll, and Abscisic Acid." *Plant Physiology* 132:779–85.

Loerch, S. 1991. "Efficacy of Plastic Pot Scrubbers as a Replacement for Roughage in High-Concentrate Cattle Diets." *Journal of Animal Science* 69:2321–28.

Lovelock, J. E. 1987. *Gaia: A New Look at Life on Earth.* Oxford: Oxford Univ. Press.

Lovins, Amory. 2001. Interview with Amory Lovins on September 8, 2001, at the Omega Institute for Holistic Studies. Interviewed by Susan Witt, E. F. Schumacher Society. Not currently available online.

Luria, S. E., and M. Delbrück. 1943. "Mutations of Bacteria from Virus Sensitivity to Virus Resistance." *Genetics* 28:491–511.

Maier, Georg. 1986. *Optik der Bilder.* Verlag der Kooperative Dürnau.

Maier, Georg, Ronald Brady, and Stephen Edelglass. 2006. *Being on Earth: Practice in Tending the Appearances.* Available online: http://natureinstitute.org/txt/gm/boe.

Makarevitch, I., S. K. Svitashev, and D. A. Somers. 2003. "Complete Sequence Analysis of Transgene Loci from Plants Transformed Via Microprojectile Bombardment." *Plant Molecular Biology* 52:421–32.

Manzanares M., H. Wada, N. Itasaki, et al. 2000. "Conservation and Elaboration of Hox Gene Regulation during Evolution of the Vertebrate Head." *Nature* 408:854–56.

Marc, J., O. Mulner-Lorillon, S. Boulben, et al. 2002. "Pesticide Roundup Provokes Cell Division Dysfunction at the Level of CDK1/Cyclin B Activation." *Chem. Res. Toxicol.* 15:326–31.

Martineau, B. 2001. *First Fruit.* New York: McGraw-Hill.

McDermott, Drew. 1981. "Artificial Intelligence Meets Natural Stupidity." In *Mind Design,* ed. John Haugeland, 143–60. Cambridge, Mass.: MIT Press.

McKibben, Bill. 2003. *Enough: Staying Human in an Engineered Age.* New York: Henry Holt.

McNab, Brian K. 1978. "Energetics of Arboreal Folivores: Physiological Problems and Ecological Consequences of Feeding on an Ubiquitous Food Supply." In *The Ecology of Arboreal Folivores,* ed. G. Gene Montgomery, 153–62. Washington, D.C.: Smithsonian Institution Press.

Mendel, Frank. 1985a. "Adaptations for Suspensory Behavior in the Limbs of Two-Toed Sloths." In *The Evolution and Ecology of Armadillos, Sloths, and Vermilinguas,* ed. G. Gene Montgomery, 151–62. Washington, D.C.: Smithsonian Institution Press.

———. 1985b. "Use of Hands and Feet of Three-Toed Sloths (*Bradypus variegatus*) during Climbing and Terrestrial Locomotion." *Journal of Mammalogy* 66:359–66.

Mendel, Frank, David Piggins, and Dale Fish. 1985. "Vision of Two-Toed Sloths (*Choloepus*)." *Journal of Mammalogy* 66:197–200.

Mendel, Gregor. 1866. *Versuch über Pflanzenhybride.* Reprint, Leipzig: Wilhelm Engelman, 1913, 1–46 (quote translated by Craig Holdrege).

Miller, H. 1999. "A Rational Approach to Labeling Biotech-Derived Foods." *Science* 284:1471–72.

Millstone, E., E. Brunner, and I. White. 1994. "Plagiarism or Protecting Public Health?" *Nature* 371:647–48.

Misawa, N., K. Masamoto, et al. 1994. "Expression of an *Erwinia* Phytoene Desaturase Gene Not Only Confers Resistance to Herbicides Interfering with Carotenoid Biosynthesis but Also Alters Xanthophyll Metabolism in Transgenic Plants." *Plant Journal* 6:481–89.

Montgomery, G. G., and M. E. Sunquist. 1975. "Impact of Sloths on Neotropical Forest Energy Flow and Nutrient Cycling." In *Tropical Ecological Systems,* ed. Frank. B. Golley and Ernesto Medina, 69–98. New York: Springer Verlag.

———. 1978. "Habitat Selection and Use by Two-Toed and Three-Toed Sloths." In *The Ecology of Arboreal Folivores,* ed. G. Gene Montgomery, 329–59. Washington, D.C.: Smithsonian Institution Press.

Moss, Lenny. 2003. *What Genes Can't Do.* Cambridge, Mass.: MIT Press.

Naples, Virginia L. 1982. "Cranial Osteology and Function in the Tree Sloths, Bradypus and Choloepus." *American Museum Novitates* 2739:1–41.

———. 1985. "The Superficial Facial Musculature in Sloths and Vermilinguas (Anteaters)." In *The Evolution and Ecology of Armadillos, Sloths, and Vermilinguas,* ed. G. Gene Montgomery, 173–89. Washington, D.C.: Smithsonian Institution Press.

Nijhout, H. F., and S. M. Paulsen. 1997. "Developmental Models and Polygenic Characters." *American Naturalist* 149:394–405.

Nord, M., M. Andrews, and S. Carlson. 2005. "Household Food Security in the United States, 2004." *Economic Research Report,* Number 11. Washington, D.C.: U.S. Department of Agriculture.

Pääbo, S. 2001. "The Human Genome and Our View of Ourselves." *Science* 291:1219–20.

Padgette, S., N. Taylor, D. Nida, et al. 1996. "The Composition of Glyphosate-Tolerant Soybean Seeds Is Equivalent to That of Conventional Soybeans." *Journal of Nutrition* 126:702–16.

Paine, J., A. Shipton, S. Chaggar, et al. 2005. "Improving the Nutritional Value of Golden Rice through Increased Pro-Vitamin A Content." *Nature Biotechnology* 23:482–87.

Pauli, George. 1999. *U.S. Regulatory Requirements for Irradiating Foods.* Washington, D.C.: Office of Premarket Approval, Food and Drug Administration. Available online: http://www.cfsan.fda.gov/~dms/opa-rdtk.html.

Peixoto, F. 2005. "Comparatives Effects of the Roundup and Glyphosate on Mitochondrial Oxidative Phosphorylation." *Chemosphere* 61:1115–22.

Pennisi, E. 2001. "The Human Genome." *Science* 291:1177–80.

———. 2007. "Working the (Gene Count) Numbers: Finally, a Firm Answer?" *Science* 316:1113.

Pingali, P., M. Hossain, and R. Gerpacio. 1997. *Asian Rice Bowls: The Returning Crisis?* New York: CAB International.

Poerschmann, J., A. Gathmann, J. Augustin, et al. 2005. "Molecular Composition of Leaves and Stems of Genetically Modified Bt and Near-Isogenic Non-Bt Maize—Characterization of Lignin Patterns." *Journal of Environmental Quality* 34:1508–18.

Pollan, Michael. 1998. "Playing God in the Garden." *New York Times Magazine,* October 25, pp. 44–51, 62–63, 82, 92.

———. 2002. "Power Steer." *New York Times Magazine,* March 31, pp. 44–51, 68, 71–72, 76.

Portin, Peter. 1993. "The Concept of the Gene: Short History and Present Status." *Quarterly Review of Biology* 68:173–223.

Portmann, Adolf. 1967. *Animal Forms and Patterns.* New York: Schocken Books.

Portugal, F., and J. Cohen. 1978. *A Century of DNA.* Cambridge, Mass.: MIT Press.

Prescott, V., P. Campbell, A. Moore, J. Mattes, et al. 2005. "Transgene Expression of a Bean ∝-Amylase Inhibitor in Peas Results in Altered Structure and Immunogenicity." *J. Agric. Food Chem.* 53:9023–30.

Pretty, J. N., C. Brett, D. Gee, et al. 2000. "An Assessment of the Total External Cost of UK Agriculture." *Agricultural Systems* 65:113–36.

Pretty, Jules. 2001. "Counting the Costs of Industrial Agriculture." *Proceedings of the Soil Association's 12th National Conference on Organic Food and Farming* (January), Chapter 4.2.

Pretty, Jules, and Rachel Hine. 2001. *Reducing Food Poverty with Sustainable Agriculture.* Colchester, UK: University of Essex.

Quiring, R., U. Walldorf, U. Kloter. 1994. "Homology of the Eyeless Gene of *Drosophila* to the Small Eye Gene in Mice and Aniridia in Humans." *Science* 265:785–89.

Ribe, Neil, and Friedrich Steinle. 2002. "Exploratory Experimentation: Goethe, Land, and Color Theory." *Physics Today* 55 (7) (July): 43–49.

Richard, S., S. Moslemi, H. Sipahutar, et al. 2005. "Differential Effects of Glyphosate and Roundup on Human Placental Cells and Aromatase." *Environmental Health Perspectives* 113:716–20.

Riegner, Mark. 1993. "Toward a Holistic Understanding of Place: Reading a Landscape through Its Flora and Fauna." In *Dwelling, Seeing and Designing: Toward a Phenomenological Ecology,* ed. David Seamon, 181–215. Albany, N.Y.: SUNY Press.

———. 1998. "Horns, Hooves, Spots, and Stripes: Form and Pattern in Mammals." In Seamon and Zajonc 1998, 177–212.

Riezler, Kurt. 1940. *Physics and Reality: Lectures of Aristotle on Modern Physics.* New Haven, Conn.: Yale Univ. Press.

Risch, N., and D. Botstein. 1996. "A Manic Depressive History." *Nature Genetics* 12:351–53.

Roessner, U., A. Luedemann, D. Brust, et al. 2001. "Metabolic Profiling Allows Comprehensive Phenotyping of Genetically or Environmentally Modified Plant Systems." *Plant Cell* 13:11–29.

Rosenberg, Susan M. 2001. "Evolving Responsively: Adaptive Mutation." *Nature Reviews Genetics* 2:504–15.

Rosenberg, Susan M., and P. J. Hastings. 2004. "Adaptive Point Mutation and Adaptive Amplification Pathways in the *Escherichia coli* Lac Systems: Stress Responses Producing Genetic Change." *Journal of Bacteriology* 186:4838–43.

Rosset, Peter. 2005. "Transgenic Crops to Address Third World Hunger? A Critical Analysis." *Bulletin of Science, Technology and Society* 25:306–13.

Rosset, Peter, and Anuradha Mittal. 2000. "There Is Food for All: Access Is the Problem." *Wall Street Journal,* letter to the editor, December 21, A19.

———. 2001. "The Paradox of Plenty." *Wall Street Journal,* letter to the editor, January 17, A27.

Roth, J. R., E. Kugelberg, A. B. Reams, et al. 2006. "Origins of Mutations under Selection: The Adaptive Mutation Controversy." *Annual Review of Microbiology* 60:477–501.

The Royal Society. 2002. "Genetically Modified Plants for Food Use and Human Health—An Update," February 4. Available online: http://www.royalsoc.ac.uk/document.asp?tip=0&id=1404.

Rozentuller, Vladislav, and Steve Talbott. 2005. "From Two Cultures to One: On the Relation Between Science and Art." Available online: http://natureinstitute.org/pub/ic/ic13/oneculture.htm.

Russell, Bertrand. 1981. *Mysticism and Logic.* Totowa, N.J.: Barnes and Noble.

Saxena, D., S. Flores, and G. Stotsky. 1999. "Insecticidal Toxin in Root Exudates from Bt Corn." *Nature* 402 (December 2): 480.

Saxena, D., and G. Stotzky. 2001. "*Bt* Corn Has a Higher Lignin Content Than Non-*Bt* Corn." *American Journal of Botany* 88:1704–6.

Schad, Wolfgang. 1977. *Man and Mammals: Toward a Biology of Form.* Garden City, N.Y.: Waldorf Press.

———, ed. 1983. *Goetheanistische Naturwissenschaft.* Vol. 3: *Zoologie.* Stuttgart: Verlag Freies Geistesleben.

Seamon, David, and Arthur Zajonc, eds. 1998. *Goethe's Way of Science: Toward a Phenomenology of Nature.* Albany, N.Y.: SUNY Press.

Serageldin, Ismail. 1999. "Biotechnology and Food Security in the 21st Century." *Science* 285:387–89.

Service, Robert. 2007. "A Growing Threat Down on the Farm." *Science* 316:1114–17.

Shannon, Claude E., and Warren Weaver. 1963. *The Mathematical Theory of Communication.* Urbana, Ill.: Univ. of Illinois Press.

Shiva, Vandana. 2000. "Genetically Engineered Vitamin A Rice: A Blind Approach to Blindness Control." Available online: http://www.gene.ch:80/genet/2000/Feb/msg00064.html.

Simon, Herbert A. 1965. *The Shape of Automation for Men and Management.* New York: Harper and Row.

Simon, Herbert A., and Allen Newell. 1958. "Heuristic Problem Solving: The Next Advance in Operations Research." *Operations Research* 6:1–10.

Singer, M., and P. Berg. 1991. *Genes and Genomes: A Changing Perspective.* Mill Valley: University Science Books.

Slijper, E. J. 1942. "Biologic-Anatomical Investigations on the Bipedal Gait and Upright Posture in Mammals, with Special Reference to a Little Goat, Born without Forelegs." *Proc. Ned. Akad. Wet.* (Amsterdam) 45:288–95, 407–15.

———. 1946. *Comparative Biologic-Anatomical Investigations on the Vertebral Column and Spinal Musculature of Mammals.* Amsterdam: N. V. Noord-Hollandsche Uitgevers Maatschappij.

Smith, Jeffrey M. 2007. *Genetic Roulette: The Documented Health Effects of Genetically Engineered Foods.* Fairfield, Iowa: Yes! Books.

Smith, Margaret. 2005. "Seeds of Discord." *Nature* 434:957–58.

Steiner, Rudolf. 1964. *The Philosophy of Freedom.* London: Rudolf Steiner Press.

———. 1988. *Goethean Science.* Spring Valley, N.Y.: Mercury Press.

Stock, G., and J. Campbell. 2000. *Engineering the Human Germline.* New York: Oxford Univ. Press.

Stocking, M. A. 2003. "Tropical Soils and Food Security: The Next 50 Years." *Science* 302:1356–58.

Strachan, J. M. N.d. "Plant Variety Protection: An Alternative to Patents." http://www.nal.usda.gov/pgdic/Probe/v2n2/plant.html.

Sturtevant, A. H., and G. W. Beadle. 1962. *An Introduction to Genetics.* 1939. Reprint, New York: Dover.

Suchantke, Andreas. 2001. *Eco-Geography.* Great Barrington, Mass.: Lindisfarne Press.

———. 2002. *Metamorphose.* Stuttgart, Germany: Verlag Freies Geistesleben.

Sugarman, Shirley, ed. 1976. *The Evolution of Consciousness.* Middletown, Conn.: Wesleyan Univ. Press.

Sunquist, Fiona. 1986. "Secret Energy of the Sloth." *International Wildlife* 16:6–10.

Sunquist, M. E., and G. G. Montgomery. 1973. "Activity Patterns and Rates of Movement of Two-Toed and Three-Toed Sloths (*Choloepus hoffmanni* and *Bradypus infuscatus*)." *Journal of Mammalogy* 54:946–54.

Tabashnik, B. E., T. J. Dennehy, Y. Carriere, et al. 2005. "Delayed Resistance to Transgenic Cotton in Pink Bollworm." *PNAS* 102:15389–93.

Talbott, Stephen L. 1995. "Can We Transcend Computation?" Chapter 23 in *The Future Does Not Compute: Transcending the Machines in Our Midst.* Sebastopol Calif.: O'Reilly and Associates. Available online: http://netfuture.org/fdnc/index.html.

Talbott, Steve. 2003a. "An Ecological Conversation with Nature." *The New Atlantis* (fall): 34–46. Available online: http://www.thenewatlantis.com/archive/3/talbott.htm.

———. 2003b. "The Vanishing World-Machine." Available online: http://qual.natureinstitute.org.

———. 2004. "Do Physical Laws Make Things Happen?" Available online: http://qual.natureinstitute.org.

Tank, W. 1984. *Tieranatomie für Künstler.* Ravensburg, Germany: Otto Maier Verlag.

Tirler, Hermann. 1966. *A Sloth in the Family.* London: Harvill Press.

USDA. 2004. *Economic Effects of U.S. Dairy Policy and Alternative Approaches to Milk Pricing,* Report to Congress. Washington, D.C.: U.S. Department of Agriculture.

van den Berg, Jan Hendrik. 1975. *The Changing Nature of Man.* New York: Dell.

Venter, J. C., M. D. Adams, E. W. Myers, et al. 2001. "The Sequence of the Human Genome." *Science* 291:1304–51.

Waage, J. K., and R. C. Best. 1985. "Arthropod Associates of Sloths." In *The Evolution and Ecology of Armadillos, Sloths, and Vermilinguas,* ed. G. Gene Montgomery, 297–322. Washington, D.C.: Smithsonian Institution Press.

Wade, Nicholas. 1998. "The Struggle to Decipher Human Genes." *New York Times,* March 10, F1 and F5.

———. 2000. "Genetic Code of Human Life Is Cracked by Scientists." *New York Times,* June 27, A1 and A21.

Walls, Laura Dassow, ed. 1999. *Material Faith: Henry David Thoreau on Science.* New York: Houghton Mifflin.

Watson, James. 1973. *The Molecular Biology of the Gene,* 3rd ed. Menlo Park, Calif.: Benjamin/Cummings.

Weinberg, Steven. 1977. *The First Three Minutes: A Modern View of the Origin of the Universe.* New York: Basic Books.

———. 1992. *Dreams of a Final Theory: The Search for the Fundamental Laws of Nature.* New York: Pantheon Books.

Weizenbaum, Joseph. 1976. *Computer Power and Human Reason.* New York: W. H. Freeman.

West-Eberhard, Mary Jane. 2003. *Developmental Plasticity and Evolution.* Oxford: Oxford Univ. Press.

Whitehead, Alfred North. 1967. *Science and the Modern World.* New York: Free Press.

Wilcove, David S., and Thomas Eisner. 2000. "The Impending Extinction of Natural History." *Chronicle of Higher Education* (September 15): 134.

Williams, G., R. Kroes, and I. Munro. 2000. "Safety Evaluation and Risk Assessment of the Herbicide Roundup and Its Active Ingredient, Glyphosate, for Humans." *Regulatory Toxicology and Pharmacology* 31:117–65.

Wilson, Alison, Jonathan Latham, and Ricarda Steinbrecher. 2004. "Gene Scrambling—Myth or Reality?" *EcoNexus Technical Report* (October). Available online: http://www.econexus.info/pdf/ENx-Genome-Scrambling-Report.pdf.

Winograd, Terry, and Fernando Flores. 1986. *Understanding Computers and Cognition: A New Foundation for Design.* Norwood, N.J.: Ablex.

Wirz, Johannes. 1998. "Progress towards Complementarity in Genetics." *Archetype* 4 (September): 21–36. Available online: http://www.ifgene.org/wirzcomp.htm.

Wolfe, Martin S. 2000. "Crop Strength through Diversity." *Nature* 406 (August 17): 681–82.

Wolpert, Lewis. 1992. *The Unnatural Nature of Science.* Cambridge, Mass.: Harvard Univ. Press.

Ye, X., S. Al-Babili, A. Klöti, et al. 2000. "Engineering the Provitamin A (Beta-Carotene) Biosynthetic Pathway into (Carotenoid-Free) Rice Endosperm." *Science* 287:303–5.

Yoon, Carol Kaesuk. 2000. "Rice Yield Doubled in Chinese Experiment without Genetic Engineering or Pesticides." *New York Times*, August 28, D1–2.

Young, J. Z. 1973. *The Life of Vertebrates,* 2nd ed. Oxford: Clarendon Press.

Zauner, Klaus-Peter, and Ehud Shapiro. 2006. Untitled essay in *Towards 2020 Science,* ed. Stephen Emmott and Stuart Rison. Cambridge, U.K.: Microsoft Research. Available online: research.microsoft.com/towards2020science.

Zhu,Y. Y., H. R. Chen, J. H. Fan, et al. 2000. "Genetic Diversity and Disease Control in Rice." *Nature* 406:718–22.

Zhu, Y. Y., Y. Y. Wang, H. R. Chen, et al. 2003. "Conserving Traditional Rice Varieties through Management for Crop Diversity." *BioScience* 53:158–62.

Index

Page numbers in italics refer to illustrations.